The Eye on Ecology: China's Ecological Monitoring Records (Volume 1)

Department of Ecology and Environment
Monitoring, Ministry of Ecology and Environment
of the People's Republic of China · China National
Environmental Monitoring Centre

The Eye on Ecology: China's Ecological Monitoring Records (Volume 1)

CEPG
Springer

Department of Ecology and Environment Monitoring, Ministry of Ecology and Environment of the People's Republic of China
Beijing, China

China National Environmental Monitoring Centre
Beijing, China

ISBN 978-981-95-0046-8 ISBN 978-981-95-0047-5 (eBook)
https://doi.org/10.1007/978-981-95-0047-5

Jointly published with China Environment Publishing Group Co., Ltd.
The print edition is not for sale in China mainland. Customers from China mainland please order the print book from: China Environment Publishing Group Co., Ltd.

This Springer imprint is published by the registered company Springer Nature Singapore Pte Ltd.
The registered company address is: 152 Beach Road, #21-01/04 Gateway East, Singapore 189721, Singapore

Foreword

Ecological environment monitoring is the foundation of ecological environment protection and provides important support for the development of ecological civilization. Since the 18th National Congress of the Communist Party of China (CPC) in 2012, the CPC Central Committee with Comrade Xi Jinping at its core has attached great importance to ecological and environmental monitoring. During his inspection tour to the Ecological and Environmental Monitoring Center of Qinghai Province in August 2016, President Xi Jinping stressed that eco-environmental protection requires thorough stock taking and ongoing monitoring of changes, and it is essential to deliver on the foundational work of building the ecological and environmental monitoring network. At its Third Plenary Session in 2024, the 20th CPC Central Committee included "improving the ecological and environmental monitoring and assessment systems" as key tasks of structural reform in ecological conservation. At the National Working Conference on Ecological and Environmental Protection, President Xi Jinping laid out systematic plans for building a Beautiful China in all respects and accelerating modernization featuring harmony between humanity and nature. This made specific requests and charted a new direction for speeding up the building of a modern ecological and environmental monitoring system and advancing the building of an air-space-land-sea integrated ecological quality monitoring network.

Ecosystem and biodiversity conservation provides the foundation for human survival and development. During the 14th Five-Year Plan period (2021–2025), the Ministry of Ecology and Environment attached great importance to building the ecological quality monitoring network and the corresponding working system, issued the Regional Ecological Quality Assessment Measures (Trial), and implemented the National Ecological Quality Supervision and Monitoring Work Plan (2023–2025) in collaboration with the Chinese Academy of Sciences to advance the building of an air-space-land-sea integrated ecological quality monitoring network. Comprehensive ecological quality monitoring stations are an integral part of the ecological quality monitoring network, constitute the vanguard of the Beautiful China initiative and ecological regulation, and act as the control points in ecological quality supervision, monitoring, and evaluation across China. Since 2023, the Ministry has organized two rounds of selection for comprehensive ecological quality monitoring stations and incorporated a total of 171 stations into the national ecological quality monitoring network, covering 31 provinces (autonomous regions and municipalities) as well as the Xinjiang Production and Construction Corps.

In our quest to care for the environment and probe the mysteries of nature, we developed a photo book series dedicated to these ecological quality monitoring stations. This book series presents with photos and texts China's expansive natural landscapes, diverse species, and fun facts about ecosystems, uncovering the fascinating world of ecological monitoring. This book, as the first volume of this photo book series, depicts a total of 55 stations with photos and texts, highlighting their special locations and geographic features, monitoring capabilities, and the typical species within their respective coverage. By publishing this book, we aim to enhance public awareness of and participation in biodiversity conservation, push forward the construction of comprehensive ecological quality monitoring stations, and contribute to the building of a denser ecological quality monitoring network in China. The ultimate goal is to advance the Beautiful China initiative and achieve modernization featuring harmony between humanity and nature.

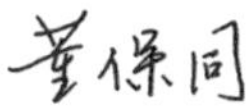

Dong Baotong
Vice Minister of Ecology
and Environment
People's Republic of China

Committees

Liu Haijiang
Hou Peng
Guo Hao

Writing Committee

Chief Editors

Dong Mingli
Gao Fengliang
Luo Haijiang
Li Mingsheng
Liu Haijiang

Associate Editors

He Jin
Huang Liutao
Wang Buying
Yu Yang
Ji Hongchao
Yang Jinjie

Members

Department of Ecological and Environmental Monitoring, Ministry of Ecology and Environment (MEE); China National Environmental Monitoring Center

Dong Guihua
Dong Mingli
Gao Fengliang
He Jin
Huang Liutao
Ji Hongchao
Kang Jing
Li Mingsheng
Li Wenjun
Liu Haijiang
Liu Haohao

Lu Sijin
Luo Haijiang
Luo Hui
Sun Cong
Sun Xiaoxuan
Wang Buying
Wang Ding
Wang Gang
Wang Shenglan
Wen Qianqian
Yang Jinjie
Yang Wu
Yu Peixin
Yu Yang
Zhang Qiming
Zhang Zijia
Zhou Jiong

Chinese Research Academy of Environmental Sciences; Satellite Application Center for Ecology and Environment, MEE; Nanjing Institute of Environmental Sciences, MEE; South China Institute of Environmental Sciences, MEE; National Marine Environmental Monitoring Center

Dong Jiahua
Feng Chaoyang
Gao Yanhua
Ge Xiaomin
Ji Wei
Jin Yuan
Liang Mingyi
Sun Chenxi
Ye Jinqing
Zhang Yuanyuan

Local (Ecological) Environmental Monitoring Centers (Stations)/Participating Organizations of Ecological Quality Monitoring Stations

Bai Lijun
Chen Junjun
Cui Qingguo
Fan Bo
Fan Ying
Guo Wei
Hu Lina
Hu Mei
Hu Zunying
Li Lei

Li Qin
Liu Jinsong
Liu Juxiu
Long Yue
Lü Xueyan
Lu Haifeng
Luo Ying
Meng Xiangliang
Pan Chengrong
Pan Shuping
Peng Tao
Qi Xiangkun
Ren Yue
Rong Zheng
Shen Qiaohui
Shi Jiankang
Su Hua
Tang Shuangli
Wang Fei
Wang Weimin
Wei Haijuan
Wei Wei
Xie Zhiyi
Xiong Jing
Xu Xun
Xu Zhuwen
Yan Ruirui
Yang Haiju
Yi Min
Yin Qiang
Yu Lizhong
Zhang Lichun
Zhang Qiang
Zhao Yongfang
Zhao Zhigang
Zhou Weiqi

Contents

Part I
Whispers of Forests

Chapter 1
Qingyuan Comprehensive Ecological Quality Monitoring Station, Liaoning Province (Forest)

Department of Ecology and Environment Monitoring, Ministry of Ecology and Environment of the People's Republic of China and China National Environmental Monitoring Centre, *The Eye on Ecology: China's Ecological Monitoring Records (Volume 1)*, https://doi.org/10.1007/978-981-95-0047-5_1

Location of the Station:

Dahu Work Zone, Changsha Village, Dasuhe Township, Qingyuan Manchu Autonomous County, Fushun City, Liaoning Province

Lead Organization:

Shenyang Institute of Applied Ecology, Chinese Academy of Sciences

Key Monitoring Area:

Changbai Mountains Water Source Conservation and Biodiversity Conservation Functional Area

Latitude and Longitude:

124° 57′ 11 ″ E

41 ° 51′ 39.6″ N

Representativeness and Typicality:

- Typical temperate secondary forest ecosystem
- Changbai Mountains Biodiversity Conservation Priority Area
- Headwaters and water source conservation functional areas of key rivers (the Hunhe River, the Qinghe River, the Chaihe River, and the Liuhe River).

Qingyuan Comprehensive Ecological Quality Monitoring Station ("Qingyuan Station" for short) is located in the northern foothills of the Longgang Mountain, the spurs of the Changbai Mountains in eastern Liaoning Province. It is characterized by a temperate, continental monsoon climate and vegetational transition between temperate coniferous and broad-leaved mixed forest and warm temperate deciduous broad-leaved forest. The zonal vegetation is predominantly broad-leaved *Pinus koraiensis* forest. This region is situated in the center of the "Northeast China Forest Belt," which is part of the "Two Barriers and Three Belts" in China's ecological security strategic pattern, playing a crucial role in water, soil, as well as biodiversity conservation.

Qingyuan Station is equipped with more than 500 sets of instruments and equipment, and has 1,350 hectares of monitoring plots targeting the typical temperate secondary forest ecosystem. This region is home to a diverse range of wildlife species, including 379 species of vascular plants in 265 genera of 92 families, such as *Taxus cuspidata*, *Pinus koraiensis*, *Phellodendron amurense*, and *Panax ginseng*. There are 143 species of vertebrates across 52 families of 21 orders, including *Rana chensinensis*, *Hynobius leechii*, and *Parus varius*.

A distant view of Qingyuan Station

Mountains in Qingyuan Station

Ker Towers in Qingyuan

Parus varius; *Sciurus vulgaris*

Picea koraiensis; *Larix kaempferi*

Adonis amurensis; *Aquilegia yabeana*

Paeonia obovata

Biodiversity Information

Taxus Cuspidata

Taxus cuspidata is an evergreen tree endemic to China in the genus *Taxus*. It is a wild plant under state priority conservation (Level I), a rare and endangered plant in China (Level I), and a national precious tree species (Level II). It is native to the Laoye Ridge, the Zhangguangcai Ridge, and Changbaishan City in Northeast China, and is also found in Japan, North Korea, and Russia.

Hynobius Leechii

Hynobius leechii is a member of the genus *Hynobius* in the family *Hynobiidae*, growing up to 14–85 mm. It is a wild animal under Level-II state priority conservation in China and listed in the IUCN Red List of Threatened Species and China's List of Wild Animals Under State Priority Conservation. This species inhabits mountainous or hilly areas at elevations ranging from 200 to 300 m, typically near small streams or inundated ponds. It generally lives on land (except during the breeding season) and is a nocturnal, hibernating animal that primarily feeds on adult insects and their larvae. It is distributed in Heilongjiang, Jilin, Liaoning, and other parts of the country.

Pinus Koraiensis

Pinus koraiensis is an evergreen coniferous tree belonging to the genus *Pinus* in the family *Pinaceae*. It can reach heights of up to 30 m and bears its perennating buds well above the surface of the ground. It is a wild plant under state priority conservation (Level II), a rare and endangered plant in China (Level II), and a national precious tree species (Level II), and is found in Liaoning, Jilin, Heilongjiang, and other parts of the country. *Pinus koraiensis* is the constructive species in the zonal vegetation of the broad-leaved *Pinus koraiensis* forest.*

* Constructive species: This term refers to the species that plays a crucial role in creating and structuring a community.

Chapter 2
Daxing'anling Comprehensive Ecological Quality Monitoring Station, Heilongjiang Province (Forest)

Department of Ecology and Environment Monitoring, Ministry of Ecology and Environment of the People's Republic of China and China National Environmental Monitoring Centre, *The Eye on Ecology: China's Ecological Monitoring Records (Volume 1)*, https://doi.org/10.1007/978-981-95-0047-5_2

Location of the Station:

Walagan Town, Tahe County, Daxing'anling Prefecture, Heilongjiang Province.

Lead Organization:

Institute of Geographic Sciences and Natural Resources Research, Chinese Academy of Sciences.

Participating Organizations:

Ecological and Environmental Monitoring Center of Heilongjiang Province.

Northeast Forestry University.

Key Monitoring Area:

Greater and Smaller Xing'an Mountains Water Source Conservation and Biodiversity Conservation Functional Area.

Latitude and Longitude:

124° 32′ E.

52° 35′ N.

Representativeness and Typicality:

- Greater Xing'an Mountains Biodiversity Conservation Priority Area
- Zonal cold temperate forest ecosystem
- Key national ecosystem service zone
- Natural ecological shield in northern and northeastern Heilongjiang

Daxing'anling Comprehensive Ecological Quality Monitoring Station ("Daxing'anling Station" for short) is located in the Greater and Smaller Xing'an Mountains Water Source Conservation and Biodiversity Conservation Functional Area, Tahe County, Daxing'anling Prefecture, Heilongjiang Province. This area is characterized by a cold temperate continental climate, and the zonal vegetation is predominantly the cold temperate forest ecosystem. The primary constructive tree species in the forests include *Larix gmelinii*, *Betula platyphylla*, and *Pinus sylvestris var. mongolica.*

Based on the Earth Critical Zone and Surface Flux Research Station of Xing'an Mountains, the Chinese Academy of Sciences, Daxing'anling Station covers an area of nearly 5,000 square meters and is equipped with 20 high-resolution, high-sensitivity monitoring instruments, including soil greenhouse gas/isotope analyzers, gas chromatographs, unmanned aerial vehicle (UAV) LiDAR, backpack LiDAR scanners, and UAV spectral imagers. This station focuses on examining the relationship between forest ecosystem structure and functioning, as well as the theoretical frameworks and methodologies for observing carbon, nitrogen, water, and other trace gas fluxes in complex terrain. Up to now, the field plots and monitoring system have been established, including large plots for biodiversity-ecosystem functioning (BEF) research, nitrogen and phosphorus addition control plots, and warming plots.

Clouds over the mountain

Aurora borealis

Among the pines; *Betula platyphylla* in the forest

Canopy gap; Shrubs

Water in the dense forest

Larix gmelinii in autumn; Wild blueberries in autumn

Chapter 3
Shanghuang Comprehensive Ecological Quality Monitoring Station, Jinhua City, Zhejiang Province (Forest)

Department of Ecology and Environment Monitoring, Ministry of Ecology and Environment of the People's Republic of China and China National Environmental Monitoring Centre, *The Eye on Ecology: China's Ecological Monitoring Records (Volume 1)*, https://doi.org/10.1007/978-981-95-0047-5_3

Location of the Station:

Damaojian Mountain, Liucheng She Ethnic Town, southwest of Wuyi County, Jinhua City, Zhejiang Province.

Lead Organization:

Institute of Atmospheric Physics, Chinese Academy of Sciences.

Participating Organizations:

Ecological and Environmental Monitoring Center of Zhejiang Province.

Zhejiang Normal University.

Key Monitoring Area:

Zhejiang-Fujian Hilly Region Biodiversity Conservation and Water Source Conservation Functional Area.

Latitude and Longitude:

119° 28.8′ E.

28° 34.8′ N.

Representativeness and Typicality

- Mixed ecosystem of subtropical evergreen broad-leaved forest, planted forest, and bamboo forest
- Key water source of the Oujiang River and boundary of the Qiantang River Basin.

Shanghuang Comprehensive Ecological Quality Monitoring Station ("Shanghuang Station" for short) sits at the top of the Damaojian Mountain in the beautiful Shanghuang Village, Wuyi County, Jinhua City, Zhejiang Province. This area is characterized by a humid subtropical monsoon climate, and the primary vegetation type is secondary coniferous and broad-leaved mixed forest.

Shanghuang Station is well equipped with high-resolution and high-sensitivity monitoring instruments, such as multi-wavelength aerosol and ozone LiDAR systems, LGR online ammonia analyzers, and carbon dioxide and methane monitoring systems. It has established 10 hectares of monitoring plots. This area boasts a rich diversity of wild plants and animals, including 1,495 species of vascular plants in 710 genera of 171 families, such as *Taxus wallichiana var. mairei*, *Anoectochilus roxburghii*, and *Euscaphis japonica*. It hosts 255 species of vertebrates across 87 families of 30 orders, including *Parus major*, *Sphenomorphus incognitus*, and *Trimeresurus stejnegeri*.

The picturesque Shanghuang Village

A sea of clouds

The scene outside Shanghuang Station

Bubulcus ibis

Nycticorax nycticorax; *Urocissa erythrorhyncha*; *Blythipicus pyrrhotis*

Leiothrix lutea; *Trimeresurus stejnegeri*

Proserpinaca palustris

Pyrrosia lingua; *Adiantum capillus-veneris*; *Begonia grandis subsp. sinensis*; *Allomorphia urophylla*

Biodiversity Information

Taxus Wallichiana Var. Mairei

Taxus wallichiana var. mairei is an evergreen tree belonging to the genus *Taxus* in the family *Taxaceae*. It can reach heights of up to 30 m and bears its perennating buds well above the ground. Endemic to China, it is a wild plant under Level-I state priority conservation, and a rare and endangered species dating back 2.5 million years to the Quaternary period, hailed as a "living fossil" of the plant kingdom. It is distributed in Shaanxi, Gansu, Anhui, Zhejiang, and other regions. All 42 countries where *Taxus wallichiana var. mairei* is found consider it a "national treasure," and the United Nations has explicitly prohibited illegal logging. Therefore, it is also known as the "giant panda" in the plant world.

Discovery of New Species

Formosania Immaculata

This new species discovered in Wuyi County, Jinhua City, belongs to the genus *Formosania* in the family *Gastromyzontidae* of the order *Cypriniformes*. With a flattened head and body and a prominent ring of barbels, this typical stream fish has only been found and recorded in the upper reaches of Xuanping Creek, a tributary of the Oujiang River in Wuyi County. This species has stringent habitat requirements, primarily inhabiting the headwaters of gravel- and cobble-bed streams.

Chapter 4
Qianjiangyuan Comprehensive Ecological Quality Monitoring Station, Zhejiang Province (Forest)

Department of Ecology and Environment Monitoring, Ministry of Ecology and Environment of the People's Republic of China and China National Environmental Monitoring Centre, *The Eye on Ecology: China's Ecological Monitoring Records (Volume 1)*, https://doi.org/10.1007/978-981-95-0047-5_4

Location of the Station

Qianjiangyuan National Forest Park, Kaihua County, Quzhou City, Zhejiang Province.

Lead Organization

Qianjiangyuan National Forest Park Administration.

Participating Organizations

Quzhou Municipal Bureau of Ecology and Environment.

Institute of Botany, Chinese Academy of Sciences.

Kaihua County Branch of Quzhou Municipal Bureau of Ecology and Environment.

Key Monitoring Area

Tianmu Mountain–Huaiyu Mountain Water Source Conservation and Biodiversity Conservation Functional Area.

Latitude and Longitude

118° 5′ E.

29° 14′ N.

Representativeness and Typicality:

- Central subtropical evergreen broad-leaved forest ecosystem at lower elevations
- Representative biodiversity and vegetation ecosystems
- The type locality for over 160 species, including *Camellia chekiangoleosa* and *Sedirea subparishii*

Qianjiangyuan Comprehensive Ecological Quality Monitoring Station ("Qianjiangyuan Station" for short) is located in Qianjiangyuan National Forest Park, Kaihua County, Quzhou City, Zhejiang Province, at the interaction of Zhejiang Province, Anhui Province, and Jiangxi Province. Qianjiangyuan National Forest Park is one of China's first 10 pilot national parks, the one and only national park in the Yangtze River Delta, and one of the 32 priority terrestrial areas for biodiversity conservation included in China's National Biodiversity Conservation Strategy and Action Plan (2011–2030).

Qianjiangyuan Station focuses on monitoring hydrological processes and functions of typical forest ecosystems, as well as comprehensive ecological quality monitoring in the region. It has built an air-space-land integrated biodiversity monitoring system, and is equipped with nearly 200 key instruments, including automated spectral imagers and hyperspectral imagers. The station has established a number of important platforms, such as a large forest plot platform, a plant diversity monitoring platform, an animal diversity monitoring platform, a forest canopy biodiversity monitoring platform, a near-ground remote sensing monitoring platform, and the Biodiversity-Ecosystem Functioning Experiment China (BEF-China) platform.

Grand Canyon Waterfall in the Grand Canyon Scenic Area, Qianjiangyuan National Forest Park

Lophura nycthemera; *Azemiops kharini*; *Muntiacus crinifrons*

Blackwellomyces kaihuaensis; *Troides aeacus*; *Emmenopterys henryi*; *Bretschneidera sinensis*; *Disanthus cercidifolius* subsp. *longipes*; *Pleione formosana*

Central subtropical evergreen broad-leaved forest at lower elevations

Biodiversity Information

Syrmaticus Ellioti

Syrmaticus ellioti is a member in the family *Phasianidae* of the order *Galliformes*, similar in size to the common pheasant. Males have a grayish-brown head, white neck, and a bright red face with bright white stripes. This ground-dwelling forest bird is highly dispersal-limited and has strong habitat dependence. Human disturbance and habitat fragmentation are the primary threats to its survival. It is included in China's List of Wild Animals Under State Priority Conservation (Level I), classified as a Near Threatened (NT) species on the IUCN Red List of Threatened Species, and also included in the Convention on International Trade in Endangered Species of Wild Fauna and Flora (CITES) Appendix I.

Discovery of New Species

Deuteragenia Ossarium

This species of spider wasp was newly discovered in Gutianshan National Nature Reserve, Kaihua County, Zhejiang Province. Females construct several chambers within their nests, placing paralyzed spiders inside to serve as food for their offspring. In the final chamber, they fill it with dead ants, forming a wall that protects the nest and offspring from parasitic wasps and predators.

Chapter 5
Dabie Mountains Comprehensive Ecological Quality Monitoring Station, Anhui Province (Forest)

Department of Ecology and Environment Monitoring, Ministry of Ecology and Environment of the People's Republic of China and China National Environmental Monitoring Centre, *The Eye on Ecology: China's Ecological Monitoring Records (Volume 1)*, https://doi.org/10.1007/978-981-95-0047-5_5

Location of the Station:

Jiefang Road, Tiantang Town, Yuexi County, Anqing City, Anhui Province.

Lead Organization:

Ecological and Environmental Monitoring Center of Anhui Province.

Participating Organizations:

Yuexi County Branch of Anqing Municipal Bureau of Ecology and Environment.

Qianshan County-level City Branch of Anqing Municipal Bureau of Ecology and Environment.

Taihu County Branch of Anqing Municipal Bureau of Ecology and Environment.

Jinzhai Branch of Lu'an Municipal Bureau of Ecology and Environment.

Huoshan County Branch of Lu'an Municipal Bureau of Ecology and Environment.

Chinese Academy of Environmental Planning.

Key Monitoring Area:

Dabie Mountains Soil and Water Conservation Functional Area.

Latitude and Longitude:

116° 21′ E.

30° 51′ N.

Representativeness and Typicality:

- Mixed forest of northern subtropical deciduous and evergreen broad-leaved trees
- Dabie Mountains Water Source Conservation and Biodiversity Conservation Key Functional Area
- Key water source conservation functional area in the middle and lower reaches of the Yangtze River and the Huaihe River.

Dabie Mountains Comprehensive Ecological Quality Monitoring Station is located in Yuexi County, Anhui Province, with five monitoring sites in Jinzhai County, Huoshan County, Qianshan County, and Taihu County. This region lies in the transition zone between the northern subtropical humid monsoon climate and the warm temperate semi-humid climate. It features complex terrain with variable slopes (typically 25° to 50°) and elevations ranging from 400 m to 1,777 m. The primary vegetation type is a northern subtropical mixed forest of deciduous and evergreen broad-leaved trees, showing clear vertical zonation patterns and offering diverse forest landscapes.

The region boasts great biodiversity, hosting over 500 vertebrate species, roughly 40 mammal species, around 172 bird species, and 1,143 insect species. Among them, animals endemic to the Dabie Mountains include *Protobothrops dabieshanensis*, *Pachyhynobius shangchengensis*, *Yerana yei*, and *Pelophylax hubeiensis*. There are 2,368 species of vascular plants, including more than 1,800 wild plant species. The endemic plants include *Pinus fenzeniana var. dabeshanensis*, *Rhododendron shanii*, and *Dendrobium huoshanense*.

Mazongling National Nature Reserve

Merops viridis; *Grus leucogeranus*; *Phoenicurus auroreus*; *Grus vipio*; *Carabus lafossei*; *Viverricula indica*

Dendrobium huoshanense

Cephalotaxus fortunei; *Euptelea pleiosperma*; *Camptotheca acuminata*; *Oyama sieboldii*; *Helwingia japonic*; *Stewartia sinensis*

Biodiversity Information

Shaniodendron Subaequal

Shaniodendron subaequal is a deciduous small tree in the genus *Shaniodendron* in the family *Hamamelidaceae*. It has exposed, small buds covered with fine hairs, and blooms in May. This species is distributed in the hilly areas of southern Yixing in Jiangsu, Anji and Lin'an in Zhejiang, the Dabie Mountains in Anhui, and the northern slopes of the Qingliang Peak in Jixi of southern Anhui. The Dabie Mountains host the largest population. This species is included in China's List of Wild Plants Under State Priority Conservation (1999 & 2021; Level I), and classified as a Critically Endangered (CR) species on the IUCN Red List of Threatened Species. It grows in the secondary mixed forests along the valleys on the lower slopes of hills.

Pinus Fenzeniana Var. Dabeshanensis

Endemic to the Dabie Mountains, *Pinus fenzeniana var. dabeshanensis* is found in Yuexi County and Jinzhai County of Anhui Province and the Dabie Mountains region in Yingshan County and Luotian County of Hubei Province. The type specimens were collected in Menkanling, Yaoluoping National Nature Reserve of Anhui, and there is a wild population in Dawangou, Yuexi County. This endangered species is included in China's List of Wild Plants Under State Priority Conservation (Level I).

Moschus Anhuiensis

Moschus anhuiensis, endemic to the Dabie Mountains, is a wild animal under Level-I state priority conservation. Its lower jaw and throat are white, with a single white stripe extending to the cheek, and there are two additional white stripes going downward along the neck, forming a ring around the upper chest. This species inhabits high mountain areas in coniferous and broad-leaved mixed forests, primarily feeding on lichens and shrub leaves. Most of them live in solitude. Due to habitat loss, continuous decline in populations, and excessive hunting, it was once critically endangered.

Discovery of New Species

Protobothrops Dabieshanensis

In 2011, the first specimen was collected in Yaoluoping National Nature Reserve, and research confirmed that this type of snake had never been recorded before. It was officially recognized as a new species in 2012, named *Protobothrops dabieshanensis.* This species belongs to the genus *Protobothrops* in the subfamily *Crotalinae* in the family *Viperidae.* It is a small to medium-sized venomous snake with tubular fangs and a slender body. Adult snakes can reach a length of about 80 cm. This species has an elongated, triangular-shaped head, distinctly separated from the neck. The head is covered with small scales; the dorsal of the head is dark brown, while the sides are yellow. A thin, dark brown stripe extends from behind the eye to the posterior border of the lower jaw. It is found only in the Dabie Mountains and is a rare, endemic snake species in China.

Chapter 6
Wuyishan Comprehensive Ecological Quality Monitoring Station, Fujian Province (Forest)

Department of Ecology and Environment Monitoring, Ministry of Ecology and Environment of the People's Republic of China and China National Environmental Monitoring Centre, *The Eye on Ecology: China's Ecological Monitoring Records (Volume 1)*, https://doi.org/10.1007/978-981-95-0047-5_6

Location of the Station:

Top of Sixin Motianling, Xingcun Town, Wuyishan County-level City, Nanping City, Fujian Province.

Lead Organization:

Fujian Academy of Environmental Sciences.

Participating Organizations:

Nanjing Institute of Environmental Sciences, Ministry of Ecology and Environment.

Ecological and Environmental Monitoring Center of Fujian Province.

Key Monitoring Area:

Zhejiang-Fujian Hilly Region Biodiversity Conservation and Water Source Conservation Functional Area.

Latitude and Longitude:

117° 43′ E.

27° 35′ N.

Representativeness and Typicality:

- Typical central subtropical forest ecosystem
- Regional biodiversity gene bank
- The world's best preserved native forest ecosystem at the same latitude
- Complete vertical gradient of vegetation in southeastern China.

Wuyishan Comprehensive Ecological Quality Monitoring Station is located on the top of Sixin Motianling, Xingcun Town, Wuyishan County-level City, Nanping City, Fujian Province, at an elevation of 1,139 m. This region belongs to the central subtropical monsoon climate zone, experiences abundant precipitation with significant vertical variation, and features a complete vertical gradient of vegetation.

The station is fully equipped with monitoring devices such as the multi-source satellite data receiving station, UAV hyperspectral cameras, LiDAR scanning system, infrared cameras, bird recorders, and vegetation-atmosphere GHG eddy covariance measurement system. This region has high biodiversity, home to 2,866 species of higher plants from 261 families, such as *Bretschneidera sinensis*, *Pseudotaxus chienii*, and *Asplenium trichomanes*. There are also 690 species of vertebrates from 148 families of 37 orders, including *Tragopan caboti*, *Capricornis milneedwardsii*, and *Pseudoxenodon karlschmidti*.

A panoramic view of Motianling

The mountain scenery of Wuyishan

Glaucidium brodiei; *Vibrissaphora liui*; *Pareas stanleyi*

Houpoea officinalis; *Selaginella wuyishanensis*; *Anoectochilus roxburghii*; *Angiopteris fokiensis*; *Gerronema*; *Eurotium parviverruculosus*

Biodiversity Information

Teinopalpus Aureus

Teinopalpus aureus, endemic to China, is famous for its unique golden spots on the wings and is the only butterfly species under Level-I state priority conservation, known as China's "national butterfly" and "crown of butterflies." It is highly sensitive to environmental changes. This species typically inhabits evergreen broad-leaved woodland at an elevation of about 1,000 m, and the host plants include *Magnolia fordiana*, *Magnolia*, and *Magnolia figo*. *Teinopalpus aureus* has five subspecies. China is home to four of them: Nominate subspecies, Guangdong subspecies, Wuyishan subspecies, and Hainan subspecies. The subspecies *T. a. shinkaii* is primarily found in Vietnam and Laos.

Discovery of New Species

Synochoneura Wuyishana

Synochoneura wuyishana is a new species first discovered in Xianfeng Ridge of Wuyishan National Park and belongs to the genus *Synochoneura Obraztsov* in the tribe *Archipini* of the subfamily *Tortricinae* and the family *Tortricidae* under the superfamily *Tortricoidea* of the order *Lepidoptera*. Currently, it has been found only in the Wuyishan region.

* Nominate subspecies: When a species is split into subspecies, the originally described population is retained as the "nominotypical subspecies" or "nominate subspecies," which repeats the same name as the species.

Discovery of New Species

Rana Wuyiensis

Rana wuyiensis was first discovered in the Tongmu Village-Masu Village area of Wuyishan National Park. This species is a member of the genus *Pseudorana*, primarily inhabiting small and shallow streams and nearby grasslands in evergreen broad-leaved forests at elevations between 700 m and 1,400 m. It has multiple rows of transverse fine skin folds on the dorsal side of its limbs. The males have a grayish-brown back densely covered with dark brown spots, while the females are reddish-brown. Currently, it is only found in the Tongmu Village-Masu Village area of Wuyishan National Park.

Chapter 7
Qianyanzhou Comprehensive Ecological Quality Monitoring Station, Jiangxi Province (Forest)

Department of Ecology and Environment Monitoring, Ministry of Ecology and Environment of the People's Republic of China and China National Environmental Monitoring Centre, *The Eye on Ecology: China's Ecological Monitoring Records (Volume 1)*, https://doi.org/10.1007/978-981-95-0047-5_7

Location of the Station:

Guanxi Town, Taihe County, Ji'an City, Jiangxi Province.

Lead Organization:

Ecological and Environmental Monitoring Center of Jiangxi Province.

Participating Organization:

Institute of Geographic Sciences and Natural Resources Research, Chinese Academy of Sciences.

Key Monitoring Areas:

Luoxiao Mountains Water Source Conservation and Biodiversity Conservation Functional Area.

Wuyishan Forest and Biodiversity Conservation Functional Area.

Latitude and Longitude:

115°04′ E.

26° 44′ N.

Representativeness and Typicality:

- Typical southern red soil hilly area
- The "Qianyanzhou model" of ecological restoration is included in geography textbooks.

Qianyanzhou Comprehensive Ecological Quality Monitoring Station ("Qianyanzhou Station" for short) is located in Guanxi Town, Taihe County, Jiangxi Province. It is a typical red soil hilly region in southern China and characterized by a typical subtropical monsoon climate. The zonal vegetation is central subtropical evergreen broad-leaved forests.

This station is well equipped and capable of conducting space-air-ground integrated ecological observations. It is among the first carbon, nitrogen, and water flux observation towers in China, established in 2002, and among the earliest field stations to conduct surface flux and earth critical zone observations and research. Focusing on ecological restoration and sustainable economic development, it pioneered the "Qianyanzhou model" characterized by the layout of "forests and grasslands on hills, ponds between hills, and orchards, fisheries, and crops in river valleys and floodplains." This model has been included in geography textbooks as a success story of coordinated ecological and economic development in red soil hilly regions. It has also been recognized by UNESCO as a case study for ecological restoration.

Building of Qianyanzhou Station

A distant view of Qianyanzhou Station

Grus leucogeranus; *Accipiter gentilis*; Larvae of *Charaxes bernardus*; *Macaca mulatta*

Viverra zibetha; *Manis pentadactyla*; *Woodwardia japonica*; *Phoebe zhennan*; *Phyllostachys violascens 'Prevernalis'*; *Ardisia crenata*; *Rubus rosifolius*

Biodiversity Information

Bretschneidera Sinensis

Bretschneidera sinensis is a deciduous tree belonging to the genus *Bretschneidera* in the family *Akaniaceae.* It is a Tertiary relict plant endemic to China, a wild plant under Level-II state priority conservation, and holds significant value for studying angiosperm phylogeny, paleogeography, and paleoclimate. When it blooms in early summer, the tree is covered in pink flowers resembling the red glow at the horizon, giving it high ornamental value. It is an excellent choice for garden landscaping and afforestation. However, due to human destruction, the scarcity of seed production, and other challenges to population regeneration, this species is currently under threat of extinction.

Camphora Micrantha

Camphora micrantha is an evergreen tree belonging to the genus *Cinnamomum* in the family *Lauraceae.* It can reach heights of up to 30 m and is a plant under Level-II state priority conservation. It is distributed in Guangxi, Guangdong, Hunan, Jiangxi, Fujian, and Taiwan, and is a disjunct plant species between Taiwan Island and the Chinese mainland. Studying this species has important scientific implications for understanding regional flora. This plant can be used to extract essential oils, serves as important industrial raw materials, and is highly suitable for papermaking. It has a well-developed root system and high water content in the trunk, making it an ideal plant for water and soil conservation.

Bretschneidera sinensis

Chapter 8
Kunyushan Comprehensive Ecological Quality Monitoring Station, Shandong Province (Forest)

Department of Ecology and Environment Monitoring, Ministry of Ecology and Environment of the People's Republic of China and China National Environmental Monitoring Centre, *The Eye on Ecology: China's Ecological Monitoring Records (Volume 1)*, https://doi.org/10.1007/978-981-95-0047-5_8

Location of the Station:

Kunyushan National Nature Reserve, Muping District, Yantai City, Shandong Province.

Lead Organizations:

Nanjing Institute of Environmental Sciences, Ministry of Ecology and Environment.

Participating Organizations:

Ecological and Environmental Monitoring Center of Yantai.
Kunyushan National Nature Reserve Administration.
Anhui Institute of Geological Surveying and Mapping.
Nanjing Forestry University.

Key Monitoring Area:

Other key ecosystem service zone.

Latitude and Longitude:

121° 46′ E.

37° 15′ N.

Representativeness and Typicality:

- Native place and natural distribution center of *Pinus densiflora* in China and Northeast Asia
- The only concentrated distribution area of the venomous snake *Gloydius intermedius* in Shandong Province.

Kunyushan Comprehensive Ecological Quality Monitoring Station ("Kunyushan Station" for short) is located in Kunyushan National Nature Reserve, Yantai City, Shandong Province, which is the only national nature reserve of forest ecosystems in Shandong. Kunyushan is the highest mountain in the Jiaodong Peninsula and known as the "ancestor of the fairy mountains on the sea," and its highest peak, Taiboding, is 923 m tall. This region is surrounded by sea on three sides and features both continental and maritime climates. It boasts the world's best-preserved *Pinus densiflora* forest ecosystem and is the only concentrated distribution area of the venomous snake *Gloydius intermedius* in Shandong.

Kunyushan Station is part of the East Shandong Hilly Ecological Zone–Kunyushan Biodiversity Conservation Functional Area. It has a 25-hectare standard plot for *Pinus densiflora* and an animal monitoring network consisting of 260 infrared cameras. It has established a comprehensive monitoring system that covers various elements including forest ecosystems, wildlife, meteorology, and soil.

Kunyushan in Shandong Province; Flowers blooming on Kunyushan; An aerial view of Kunyushan Station

Parus major; *Aegithalos caudatus*; *Dendrocopos major*; *Lanius bucephalus bucephalus*; *Hypsipetes amaurotis*

Thymelicus sylvaticus; *Argynnis paphia*; *Parena latecincta*; *Lelia decempunctata*; *Aegeriidae sesiidae*; *Lobelia sessilifolia*; *Nymphoides coreana*; *Nabalus tatarinowii*

Biodiversity Information

Pinus Densiflora

Pinus densiflora is an evergreen tree belonging to the genus *Pinus* in the family *Pinaceae*, and is named for its reddish-brown bark. It is native to China, Japan, and Russia, and Kunyushan in Shandong Province is the native place and natural distribution center of *Pinus densiflora* in China. In 2008, the State Council approved the establishment of Kunyushan National Nature Reserve, the only national nature reserve dedicated to the conservation of *Pinus densiflora*. This species can be planted in courtyards for ornamental purposes. Its trunk can be tapped to extract resin, which can be processed into rosin and turpentine for medicinal use. The wood has a straight grain and is hard and highly resistant to decay, making it suitable for producing railway ties, furniture, and other products. The seeds can be used to produce oil for both culinary and industrial use. The needles can be used to extract essential oils. It is a wind-resistant plant and serves as a primary afforestation species in coastal mountainous regions including the Liaodong Peninsula.

Pinus densiflora cones

Pinus densiflora inflorescence; *Pinus densiflora*

Species Endemic to Shandong

Anemone Shikokiana

Anemone shikokiana is a perennial herb belonging to the genus *Anemone* in the family *Ranunculaceae*. It can grow up to 55 cm tall and flowers from June to July. This species is primarily found in mountainous grasslands at elevations of 600 to 1,100 m in eastern Shandong including Laoshan District of Qingdao City and Muping District of Yantai City. It has medicinal properties that help cool the body, reduce bleeding, and dispel dampness.

Chapter 9
Enshi Comprehensive Ecological Quality Monitoring Station, Hubei Province (Forest)

Department of Ecology and Environment Monitoring, Ministry of Ecology and Environment of the People's Republic of China and China National Environmental Monitoring Centre, *The Eye on Ecology: China's Ecological Monitoring Records (Volume 1)*, https://doi.org/10.1007/978-981-95-0047-5_9

Location of the Station:

Enshi Tujia and Miao Autonomous Prefecture, Hubei Province.

Lead Organization:

Ecological and Environmental Monitoring Center of Hubei Province.

Participating Organizations:

Hubei University.

Hubei Qizimeishan National Nature Reserve Administration.

Key Monitoring Area:

Wuling Mountains Biodiversity and Soil and Water Conservation Functional Area.

Latitude and Longitude:

109° 28′ E.

29° 59′ N.

Representativeness and Typicality:

- The core area of the East Sichuan-West Hubei Endemism Distribution Center
- The core area of central subtropical flora distribution
- Forest ecosystem dominated by primitive *Davidia involucrata* community
- Subalpine *Sphagnum palustre* swamp.

Enshi Comprehensive Ecological Quality Monitoring Station ("Enshi Station" for short) is located in Qizimeishan National Nature Reserve, Xuan'en County, in the Wuling Mountains area of southwestern Hubei Province. This region lies in the watershed between the Qingjiang River, an important tributary of the middle reaches of the Yangtze River, and the Yuanjiang River, the largest tributary of the Dongting Lake system. It hosts forest ecosystems dominated by primitive *Davidia involucrata* communities, subalpine *Sphagnum palustre* swamp ecosystems, and habitats for rare and precious species like *Moschus berezovskii*. This station boasts complete monitoring infrastructure and experiment platforms, as well as a comprehensive biodiversity monitoring system and application support platform. It is equipped with infrared cameras and fully automated integrated monitoring video equipment, and has set up 26 fixed field monitoring plots. The station primarily undertakes tasks including soil and water conservation, water source conservation, biodiversity conservation, ecological quality monitoring, and regional ecological assessments in the Wuling Mountains area. It also conducts nature education programs.

Autumn scenery of Qizimeishan National Nature Reserve; Broad-leaved tree communities; *Sphagnum palustre* wetland

Petaurista alborufus

Chrysolophus pictus; *Paguma larvata*; *Muntiacus vaginalis*

Troides aeacus

Taxus wallichiana var. mairei; *Hosta ventricosa*

Biodiversity Information

Davidia Involucrata

Davidia involucrata, also known as the dove tree, is a deciduous tree belonging to the genus *Davidia* in the family *Nyssaceae*. During its peak flowering period, two large white bracts beneath each head-like inflorescence are particularly prominent. Endemic to China, it is a wild plant under Level-I state priority conservation and a Tertiary relict plant, hailed as a "living fossil of the plant kingdom." Primitive *Davidia involucrata* communities are distributed on both sides of the shady and humid ravines at elevations between 1,200 and 1,700 m in the Qizimeishan and Badagongshan mountains. The communities feature abundant genetic resources and a stable ecosystem structure, offering valuable research samples for global ecological conservation.

Sphagnum Palustre

Sphagnum palustre is characterized by its soft, grayish-white or grayish-green plant body, which can reach heights of up to dozens of centimeters. The upper part of the plant continues to grow upward, while the lower part dies and contributes to the formation of peat, which stores nearly half of the world's peatland carbon. This species has a remarkable ability to absorb and retain water, holding 10 to 25 times its dry weight in moisture. There are well-preserved *Sphagnum palustre* swamps distributed in a patchy pattern in the area northeast of Enshi Station, at elevations between 1,650 and 1,950 m.

Subalpine *Sphagnum palustre* swamp, which is rare in Central China

Chapter 10
Shennongjia Comprehensive Ecological Quality Monitoring Station, Hubei Province (Forest)

Department of Ecology and Environment Monitoring, Ministry of Ecology and Environment of the People's Republic of China and China National Environmental Monitoring Centre, *The Eye on Ecology: China's Ecological Monitoring Records (Volume 1)*, https://doi.org/10.1007/978-981-95-0047-5_10

Location of the Station:

Xingshan County, Yichang City, Hubei Province.

Lead Organization:

Institute of Botany, Chinese Academy of Sciences.

Participating Organization:

Ecological and Environmental Monitoring Center of Hubei Province.

Key Monitoring Area:

Daba Mountains Biodiversity Conservation Functional Area.

Latitude and Longitude:

110°03′ E.

31° 19′ N.

Representativeness and Typicality:

- Core area of the East Sichuan-West Hubei Endemism Distribution Center
- Complete vertical vegetation spectrum based on the northern subtropical forest ecosystem
- One of the areas with the richest biodiversity in the northern subtropical region of China.

Shennongjia Comprehensive Ecological Quality Monitoring Station ("Shennongjia Station" for short) is located in Xingshan County, Yichang City, on the southern slope of the Shennongjia region, known as the "Roof of Central China." Located at an elevation of 1,290 m, it lies within the Qinling-Daba Mountains key functional area of biodiversity conservation and water source conservation. This area features a northern subtropical humid monsoon climate and a complete vertical vegetation spectrum based on the north subtropical forest ecosystem. This area is critical for the differentiation, development, and concentration of the temperate flora and leads the world with diverse woody plant species. It serves as a refuge for numerous ancient relict species and also an important habitat for the world's rare and endangered species and species endemic to China.

Shennongjia Station has comprehensive infrastructure, including a meteorological observation field, a botanical garden of rare and endangered plants, experimental greenhouses, field observation sites, and a forest vertical spectrum research tower. The station houses over 15,000 herbarium specimens and more than 36,000 soil samples.

An aerial view of Shennongjia Station

Evergreen broad-leaved forest; Coniferous and broad-leaf mixed forest

Subalpine coniferous forest; Deciduous broad-leaved forest; Mixed forest of evergreen deciduous and broad-leaved trees in Shennongjia; Davidia involucrata; *Thuja sutchuenensis*; *Rhinopithecus roxellana*

Biodiversity Information

Acer Griseum

Acer griseum, a tall deciduous tree belonging to the genus *Acer* in the family *Aceraceae*, is a rare and precious plant endemic to China. It can reach up to 30 m, with a trunk diameter exceeding 1 m and a crown diameter of nearly 20 m. *Acer griseum* is renowned primarily for its distinct ochre-brown or flesh-red bark. The papery bark peels off in curls, revealing smooth red trunks, which creates a special landscape in the forest. This species is mainly distributed in southwestern Henan, southern Shaanxi, western Hubei, and eastern Sichuan, typically found in sparse forests at elevations between 1,500 and 2,000 m.

Over a hundred years ago, British plant collector and explorer Ernest Henry Wilson discovered *Acer griseum* in Xingshan County, Hubei Province, and later, introduced it to Wales and several other countries. The tree planted at the Arnold Arboretum at Harvard University has now grown into a towering tree, and the university held a centennial celebration for this tree in 2007.

Acer Griseum

Acer Griseum

Chapter 11
Mangshan Comprehensive Ecological Quality Monitoring Station, Hunan Province (Forest)

Department of Ecology and Environment Monitoring, Ministry of Ecology and Environment of the People's Republic of China and China National Environmental Monitoring Centre, *The Eye on Ecology: China's Ecological Monitoring Records (Volume 1)*, https://doi.org/10.1007/978-981-95-0047-5_11

Location of the Station:

Chawang Valley, Mangshan National Forest Park, Yizhang County, Chenzhou City, Hunan Province.

Lead Organization:

Ecological and Environmental Monitoring Center of Chenzhou City.

Participating Organizations:

Chenzhou Mangshan State-owned Forestry Administration.

Research Institute of Tropical Forestry, Chinese Academy of Forestry.

Key Monitoring Area:

Nanling Mountains Forest and Biodiversity Conservation Functional Area.

Latitude and Longitude:

112° 54′ E.

24° 58′ N.

Representativeness and Typicality

- Core area of the Nanling Mountains, a climate boundary between northern and southern China
- Transition zone between central subtropical climate and southern subtropical climate
- Globally recognized biodiversity hotspot
- Watershed between the Yangtze River Basin and the Pearl River Basin, and source of the Beijiang River, a primary tributary of the Pearl River.

Mangshan Comprehensive Ecological Quality Monitoring Station ("Mangshan Station" for short) is located in Mangshan National Nature Reserve, which is also the largest forest park in Hunan. It sits in the core area of the Nanling Mountains, a climate boundary between northern and southern China, is a transition zone between central subtropical climate and southern subtropical climate, and serves as a watershed between the Yangtze River Basin and the Pearl River Basin.

Mangshan Station has nearly 1,500 square meters of functional rooms and facilities and equipment such as monitoring plots, hydrological monitoring stations, a monitoring network of infrared cameras, and carbon flux systems. This region has great biodiversity, hosting 2,404 species of vascular plants in 970 genera of 213 families and 374 species of vertebrates across five classes, 33 orders, and 108 families. There are 58 wild plant species under state priority conservation. Among them, *Taxus wallichiana var. mairei* and *Dendrobium flexicaule* are under Level-I state priority conservation, and 56 species are under Level-II state priority conservation. A total of 73 wild animal species have been recorded, including 11 species under Level-I state priority conservation, such as *Ermia mangshanensis*, *Tragopan caboti*, and *Syrmaticus ellioti*, and 62 species under Level-II state priority conservation.

The amphibian transect of Mangshan Station

Pleione bulbocodioides; *Taxus wallichiana var. mairei*; *Coptis chinensis*; *Bretschneidera sinensis*; *Leiothrix lutea*; *Macaca thibetana*; *Lophura nycthemera*

Pinus Kwangtungensis

Biodiversity Information

Protobothrops Mangshanensis

Protobothrops mangshanensis, a large, rear-fanged venomous snake endemic to China, belongs to the genus *Protobothrops* in the order *Squamata* of the class *Reptilia*. This snake can reach about 2 m in length, and adults generally weigh between 3 and 10 kg. It has been exclusively found in the Mangshan Mountain region of the Nanling Mountains spanning Hunan Province and Guangdong Province, generally at elevations between 600 m and 1,500 m. This species currently has a small population of around 500. Given its narrow geographic distribution and unknown genetic diversity, there is an urgent need to protect this species. It was included in CITES Appendix II in March 2013, and added to China's List of Wild Animals Under State Priority Conservation (Level I) in February 2021, classified as one of the 48 species that urgently require protection efforts to prevent extinction.

Discovery of New Species

Tylototriton Asperrimus

Tylototriton asperrimus, a new species discovered in Mangshan Mountain, belongs to the genus *Tylototriton* in the family *Salamandridae* of the order *Urodela*. It is a wild animal species under Level-II state priority conservation. This species is characterized by its elongated body and flattened head with an indentation on the top, and the length of the head is greater than its width. It has only been found in some parts of the eastern Mangshan forest region and in the Nanling Mountains area, including two villages in Shaping Town of Lechang City in Guangdong, which border Ruyuan County of Guangdong and Yizhang County of Hunan. Given its limited wild population, narrow geographic distribution, and over-capturing, this species was included in CITES Appendix II in 2019.

Chapter 12
Zhangjiajie Comprehensive Ecological Quality Monitoring Station, Hunan Province (Forest)

Department of Ecology and Environment Monitoring, Ministry of Ecology and Environment of the People's Republic of China and China National Environmental Monitoring Centre, *The Eye on Ecology: China's Ecological Monitoring Records (Volume 1)*, https://doi.org/10.1007/978-981-95-0047-5_12

Location of the Station

Zhangjiajie City, Hunan Province.

Lead Organization

Ecological and Environmental Monitoring Center of Zhangjiajie City.

Participating Organizations

Wuhan Botanical Garden, Chinese Academy of Sciences.

Hunan Badagongshan National Nature Reserve Administration.

Key Monitoring Area

Wuling Mountains Biodiversity and Soil and Water Conservation Functional Area.

Latitude and Longitude

110° 5′ E.

29° 46′ N.

Representativeness and Typicality

- Well-preserved mixed forest ecosystem of evergreen deciduous and broad-leaved trees.
- One of the world's 17 Key Biodiversity Areas (KBAs).

TowerZhangjiajie Comprehensive Ecological Quality Monitoring Station ("Zhangjiajie Station" for short) is located in the middle of the Wuling Mountains, within Badagongshan National Nature Reserve. It features a humid central subtropical climate subject to mountainous monsoon and is home to a rich range of wildlife resources. Its flora and fauna include both ancient species endemic to China and modern species across the northern and southern regions.

This station has established forest ecosystem monitoring plots, a biodiversity exhibition hall, a comprehensive analysis laboratory, and a monitoring network of infrared cameras. A total of 559 species across 470 genera of 150 families have been monitored. Plant species under Level-I state priority conservation include *Davidia involucrata*, *Taxus wallichiana var. chinensis*, under Level II state priority conservation include *Bretschneidera sinensis* and *Fagus longipetiolata*, and with extremely small populations include *Changiostyrax dolichocarpus* and *Manglietia patungensis*. There are 16 animal species under state priority conservation, of which *Capricornis milneedwardsii* and *Moschus berezovskii* are under Level-I state priority conservation.

Badagongshan Observation Tower

Landscape around Zhangjiajie Station

Sciurotamias davidianus; *Mustela sibirica*; *Euprepiophis mandarinus*; *Tylototriton asperrimus*

Aviceda leuphotes; *Glaucidium cuculoides*; *Sinosuthora webbiana*

Impatiens balsamina; *Mycena pura*; *Ganoderma sichuanense*; *Hericium erinaceum*; *Russula rosacea*

Biodiversity Information

Moschus Berezovskii

Moschus berezovskii is the smallest species in the genus *Moschus*. The upper canines of the male musk deer are well-developed, elongated, and sharp, protruding out of their mouths like tusks. The musk produced by the male has high medicinal value and serves as a precious, natural, high-grade perfumery ingredient. It is a traditional export earner, known as "soft gold." The high price of musk has led to excessive hunting, resulting in a sharp decrease in wild populations and a shortage of this key ingredient in many valuable traditional Chinese medicine formulas.

Chapter 13
Dinghushan Comprehensive Ecological Quality Monitoring Station, Guangdong Province (Forest)

Department of Ecology and Environment Monitoring, Ministry of Ecology and Environment of the People's Republic of China and China National Environmental Monitoring Centre, *The Eye on Ecology: China's Ecological Monitoring Records (Volume 1)*, https://doi.org/10.1007/978-981-95-0047-5_13

Location of the Station

Zhaoqing City, Guangdong Province.

Lead Organization

South China Botanical Garden, Chinese Academy of Sciences.

Participating Organizations

Ecological and Environmental Monitoring Center of Guangdong Province.

Ecological and Environmental Monitoring Center of Shaoguan City.

Key Monitoring Area

Other key ecosystem service zone.

Latitude and Longitude

112° 32′ E.

23° 10′ N.

Representativeness and Typicality

- Evergreen coniferous forest, mixed coniferous and broad-leaved forest, and complete horizontal zonation of evergreen broad-leaved forest
- Unique southern subtropical forest at the same latitude.

Dinghushan Comprehensive Ecological Quality Monitoring Station ("Dinghushan Station" for short) consists of a main station in Zhaoqing City of Guangdong Province and the Nanling Station in Tianjingshan Forest Park of Shaoguan City of Guangdong Province. This region is located in the southern mountainous and hilly area, characterized by a typical humid subtropical monsoon climate. Horizontally, it is distributed with evergreen coniferous forests, mixed coniferous and broad-leaved forests, and evergreen broad-leaved forests. Vertically, there are mountain dwarf forests, mountain shrubs/grasses, and mountain meadows, among other vegetation types.

The station is equipped with automated meteorological radiation observation facilities, a flux observation tower for the atmosphere-forest ecosystem, automated ambient air observation facilities, runoff observation fields for forest catchments, large runoff observation fields, large forest sample plots, and permanent forest sample plots. It has established an air-space-land integrated observation platform. A total of 1,861 species of wild higher plants have been monitored, including 141 species of bryophytes, 131 species of ferns, 23 species of gymnosperms, and 1,566 species of angiosperms. As for animals, 1,125 species have been recorded, comprising 43 species of mammals, 267 species of birds, 23 species of amphibians, 54 species of reptiles, 25 species of fish, and 713 species of insects. In 2023, two new bird species in Guangdong Province were discovered for the first time: *Treron curvirostra* and *Treron sphenurus*.

A panoramic view of the Jilong Peak

A distant view of Dinghushan Station; Scenery of the Dinghu Lake

Calotes versicolor; *Lophura nycthemera*; *Trimeresurus albolabris*

Begonia fimbristipula; *Hedyotis effusa*; *Archidendron turgidum*; *Clematis tinghuensis*; *Callicarpa tingwuensis*

Biodiversity Information

Gymnosphaera Podophylla

Gymnosphaera podophylla belongs to the genus *Cyathea* in the family *Cyatheaceae*. It can grow to a height of 1 to 3 m, with large leaves measuring 2 to 3 m in length. The trunk can reach 6 m or above in height and 10 to 20 cm in diameter. This species thrives in mountain forests and along stream banks at elevations of 95 to 1,100 m, and is distributed in Taiwan, Fujian, Guangdong, and other parts of China. The relict plant is included in CITES Appendix II (2013 version) and China's List of Wild Plants Under State Priority Conservation (Level II).

Discovery of New Species

Lindera Chunii Merr

In May 1928, Chun Woon-young, a professor of botany at Sun Yat-sen University, collected specimens of the species during a botanical survey in Dinghu Mountain (or "Dinghushan"), and it was later identified as a new species. The type specimens of this species are preserved at the Herbarium of the Arnold Arboretum at Harvard University, the South China Botanical Garden Herbarium of the Chinese Academy of Sciences, and the University of California Herbarium. This species belongs to the genus *Lindera* in the family *Lauraceae* and is an evergreen small tree or shrub. Its branches are slender and initially covered with hair, which falls off over time. The flowering period lasts from February to March, and the fruiting period occurs from August to September. It is distributed in natural forests at lower elevations in Guangdong, Hainan, Guangxi, and other regions of China. Its roots are used for medicinal purposes. Due to its attractive leaf shape and the golden yellow color undersides, it also serves as an ornamental tree.

Discovery of New Species

Treron Curvirostra and Treron Sphenurus

The two species both belong to the genus *Treron* in the family *Columbidae.* They are newly recorded bird species in Dinghu Mountain and Guangdong Province, and wild animal species under Level-II state priority conservation. *Treron curvirostra* is characterized by its broad yellow edges on the outer coverts. Its short, thick bill is pale yellowish-green, with coral red on both sides of the base. *Treron sphenurus* has distinct purple-red patches on its shoulders, tawny yellow undertail coverts, and a yellowish-green abdomen. Its tail is olive green and shaped like a wedge. Both are resident birds that do not migrate.

Treron curvirostra male; *Treron sphenurus* male; *Treron sphenurus* and *Treron curvirostra* females

Chapter 14
Lijiang Comprehensive Ecological Quality Monitoring Station, Guilin City, Guangxi (Forest)

Department of Ecology and Environment Monitoring, Ministry of Ecology and Environment of the People's Republic of China and China National Environmental Monitoring Centre, *The Eye on Ecology: China's Ecological Monitoring Records (Volume 1)*, https://doi.org/10.1007/978-981-95-0047-5_14

Location of the Station

Shuangtan Village, Yangshuo Town, Yangshuo County, Guilin City, Guangxi Zhuang Autonomous Region.

Lead Organization

Ecological and Environmental Monitoring Center of Guangxi.

Participating Organization

Ecological and Environmental Monitoring Center of Guilin City.

Key Monitoring Areas

Dayao Mountains Biodiversity Conservation Functional Area.

Nanling Mountains Forest and Biodiversity Conservation Functional Area.

Latitude and Longitude

110° 30′ E.

24° 47′ N.

Representativeness and Typicality

- Typical subtropical evergreen broad-leaved forests, mixed evergreen broad-leaved forests, and planted forests
- Typical subtropical karst landform.

Lijiang Comprehensive Ecological Quality Monitoring Station ("Lijiang Station" for short) is located in Yangshuo County, Guilin City, on the bank of the renowned Lijiang River. This region is characterized by typical karst peak forest landforms and a humid subtropical monsoon climate.

This station has comprehensive infrastructure including an automated water environment monitoring station, an automated ambient air monitoring station, an atmosphere-vegetation carbon flux observation system, and an automated weather station. It is also equipped with DOAS online ammonia analyzers and environmental DNA analyzers. This region is rich in wildlife resources, hosting 3,120 species of vascular plants in 1,103 genera of 249 families. There are four plant species under Level-I state priority conservation, including *Cathaya argyrophylla, Glyptostrobus pensilis*, and *Taxus wallichiana var. mairei*, and 51 plant species under Level-II state priority conservation, such as *Alsophila spinulosa, Pinus kwangtungensis, Chamaecyparis hodginsii, Liriodendron chinense, Emmenopterys henryi, Keteleeria calcarea*, and *Cibotium barometz*. There are 95 species of fish, 10 species of benthic invertebrates, 167 species of birds, 35 species of amphibians, and 142 species of butterflies.

Forest Carbon Sink Observation Tower

Lophura nycthemera; *Pachytriton inexpectatus*

Mergus squamatus; *Tragopan temminckii*

Biodiversity Information

Hynobius Maoershanensis

Hynobius maoershanensis is a tailed amphibian belonging to the genus *Hynobius* in the family *Hynobiidae*. The total length of males is 152.3 to 160.2 mm, while females range from 136.1 to 155.2 mm. The head is relatively large and slightly flattened, with a length greater than the width. They live in mountain swamps and surrounding areas at elevations between 1,978 and 2,015 m, and the vegetation around is predominantly composed of *Tsuga chinensis* and montane dwarf forests. They prefer to eat small aquatic invertebrates such as earthworms and *Helicopsyche* larvae. It is mainly distributed in Maoershan located on the border of Longsheng County and Xing'an County of Guangxi.

Discovery of New Species

Ecological surveys in Guangxi in recent years have discovered 48 new species, with the number of newly discovered plant species ranking second nationwide during the same period. Among the newly discovered species, there are nine terrestrial wildlife species, including four amphibian species and five reptile species. A total of 39 new plant species have been discovered, including three species of lycophytes and ferns from three genera of two families and 36 species of seed plants from 25 genera of

21 families. These newly discovered species further enrich the background biodiversity in China and expand the horizons for biodiversity research. The Lijiang River Basin hosts the newly discovered plant species of *Prunus quanzhouensis* and *Isodon maoershanensis R.H.Jiang & Jian L.Li.* and the newly discovered animal species of *Nidirana guibeiensis.* Additionally, *Mergus squamatus* and *Ciconia boyciana*, both animal species under Level-I state priority conservation, have been monitored, with their population size setting new records. The discovery of these new species and the recorded numbers of active populations of rare species fully demonstrate that the Lijiang River Basin is an important gene bank for wildlife and holds significant value for biodiversity conservation.

Chapter 15
Huanjiang Karst Comprehensive Ecological Quality Monitoring Station, Guangxi (Forest)

Department of Ecology and Environment Monitoring, Ministry of Ecology and Environment of the People's Republic of China and China National Environmental Monitoring Centre, *The Eye on Ecology: China's Ecological Monitoring Records (Volume 1)*, https://doi.org/10.1007/978-981-95-0047-5_15

Location of the Station

Dacai Township, Huanjiang Maonan Autonomous County, Hechi City, Guangxi.

Lead Organization

Institute of Subtropical Agriculture, Chinese Academy of Sciences.

Participating Organizations

Ecological and Environmental Monitoring Center of Guangxi.

Ecological and Environmental Monitoring Center of Hechi City.

Key Monitoring Area

Wuling Mountains Biodiversity and Soil and Water Conservation Functional Area.

Latitude and Longitude

108° 18′ E.

24° 43′ N.

Representativeness and Typicality

- Typical karst peak-cluster depression ecosystem
- Water source conservation area in the upper reaches of the Xijiang River
- Karst soil conservation area.

Huanjiang Karst Comprehensive Ecological Quality Monitoring Station ("Huanjiang Karst Station" for short) is located in Dacai Township, Huanjiang Maonan Autonomous County, Hechi City, Guangxi. The total floor space of the field monitoring station is nearly 5,000 square meters. This region features a humid central subtropical climate and a typical karst peak-cluster depression ecosystem. The ecosystem is fragile, with high sensitivity to soil erosion, and the soil is prone to erosion, adding difficulty to ecological restoration.

The station consists of a comprehensive research demonstration park and a top plant community pilot demonstration zone. It has a ground-based ecological monitoring network covering small watersheds, transects, and regions, and is equipped with slope runoff plots, underground-aboveground soil and water monitoring facilities for karst areas, and facilities for hydrochemical monitoring, nitrogen deposition monitoring, and forest ecosystem flux observation. Additionally, it has established 25-hectare dynamic forest monitoring plots, long-term monitoring plots for small watersheds, long-term monitoring plots for vegetation restoration and ecological processes, and monitoring and control plots for typical rocky desertification control project areas.

Underground-aboveground observation site

Karst mixed forest of evergreen deciduous and broad-leaved trees

Biodiversity Information

Woonyoungia Septentrionalis

Woonyoungia septentrionalis is an evergreen tree from the family *Magnoliaceae* and under Level-I state priority conservation. The plant has alternate leaves with smooth edges, and the flowers are solitary at the tips of the branches. It is monoecious and has long stalks. Some parts of the flower split along the dorsal suture, and the seeds have arils. There are two species in the genus *Kmeria duperreana (Pierre) Dandy*, including one in China. It grows in limestone mountain forests at elevations of 300 to 500 m in northern Guangxi and southeastern Guizhou. The tree has a straight trunk, excellent wood quality, beautiful tree shape, and large, colorful flowers. It is a valuable timber species and a great ornamental tree for yard landscaping.

Discovery of New Species

Cleisostoma Mulunense

Cleisostoma mulunense is a new species belonging to the genus *Cleisostoma*, discovered in July 2023 during a survey in Guangxi Mulun National Nature Reserve. It was later identified as a new species in the family *Orchidaceae* and named *Cleisostoma mulunense.*

It has a short stem, flattened linear leaves, and an acute apex.

Chapter 16
Hainan Tropical Rainforest Station (Forest)

Department of Ecology and Environment Monitoring, Ministry of Ecology and Environment of the People's Republic of China and China National Environmental Monitoring Centre, *The Eye on Ecology: China's Ecological Monitoring Records (Volume 1)*, https://doi.org/10.1007/978-981-95-0047-5_16

Location of the Station

Baoting Li and Miao Autonomous County, Hainan Province.

Lead Organization

Ecological and Environmental Monitoring Center of Hainan Province.

Participating Organizations

Hainan University.

Hainan Academy of Environmental Sciences.

Experimental Station of the Research Institute of Tropical Forestry, Chinese Academy of Forestry.

Key Monitoring Area

Tropical rainforest area in the central mountainous area of Hainan Island.

Latitude and Longitude

109° 42′ E.

18° 38′ N.

Representativeness and Typicality

- Typical island tropical rainforest ecosystem
- Globally recognized biodiversity hotspot
- Key water source conservation area on Hainan Island.

Hainan Tropical Rainforest Station is located in Hainan Tropical Rainforest National Park and consists of Baoting Station, Xinglong Station, Wuzhishan Station, and Jianfengling Station. The region lies within a tropical marine monsoon climate zone and boasts the most diverse and best-preserved tropical rainforest on any continental island in China.

The station possesses a full range of monitoring facilities, including eddy covariance flux systems, soil flux monitoring systems, forest climate gradient systems, and forest water balance systems. It also features the world's largest forest observation plots (one 60-hectare plot and one 64-hectare plot), as well as 150 multi-square-kilometer grid plots. This region is rich in wildlife resources. It is home to 4,367 species of vascular plants from 1,368 genera of 289 families, including 149 species under state priority conservation. Among them, six are under Level-I state priority conservation, including *Hopea hainanensis*, *Paphiopedilum purpuratum*, and *Cycas hainanensis*. There are 651 wild animal species, including 134 species under state priority conservation. Among them, 14 are under Level-I state priority conservation, including *Nomascus hainanus*, *Rucervus eldii*, and *Arborophila ardens*.

Wuzhishan; Bawangling; Diaoluoshan

Rucervus eldii; *Xenopeltis hainanensis*; *Tropidophorus hainanus*; *Tylototriton hainanensis*; *Goniurosaurus zhoui*

Milvus migrans; *Chirixalus doriae*; *Draco maculatus*; *Rhacophorus rhodopus*; *Polypedates dennysi*

Hopea hainanensis; *Paphiopedilum purpuratum*

Discovery of New Species

Thismia Jinzun

Thismia jinzun is a new species discovered in Jianfengling of Hainan. It is the ninth species of the rare genus *Thismia* discovered and recorded in China and the third recorded in Hainan. It has a golden flower, a vermicular root, and an upright white stem. The floral bracts have filamentous appendages about 1 centimeter long. It is a small, fully mycoheterotrophic herb that has extremely stringent habitat requirements.

Chapter 17
Jinyunshan Comprehensive Ecological Quality Monitoring Station, Chongqing (Forest)

Department of Ecology and Environment Monitoring, Ministry of Ecology and Environment of the People's Republic of China and China National Environmental Monitoring Centre, *The Eye on Ecology: China's Ecological Monitoring Records (Volume 1)*, https://doi.org/10.1007/978-981-95-0047-5_17

Location of the Station

Core area of Jinyunshan National Nature Reserve in Chongqing.

Lead Organization

Big Data Application Center of Chongqing Municipal Bureau of Ecology and Environment.

Participating Organizations

Ecological and Environmental Monitoring Center of Chongqing Municipality.

Satellite Application Center for Ecology and Environment, Ministry of Ecology and Environment.

Chongqing Institute of Green and Intelligent Technology, Chinese Academy of Sciences.

Chongqing Beibei District Bureau of Ecology and Environment.

Key Monitoring Area

Other key ecosystem service zone.

Latitude and Longitude

106° 24′ E.

29° 50′ N.

Representativeness and Typicality

- The most representative subtropical evergreen broad-leaved forest ecosystem at the same latitude in China
- Key functional area of biodiversity conservation.

Jinyunshan Comprehensive Ecological Quality Monitoring Station ("Jinyunshan Station" for short) is located in Jinyunshan National Nature Reserve in Chongqing. It is one of the regions with well-preserved subtropical evergreen broad-leaved forest ecosystems at the same latitude in China, and has a humid subtropical monsoon climate.

The station is equipped with a comprehensive array of monitoring devices, including those for vegetation, soil, shortwave radiation, runoff, infrared thermography, airborne hyperspectral imaging, backpack LiDAR, and drones. It also has sample plots for various types of vegetation. This region boasts a high diversity of plants, including some old-known plant species and plenty of rare and endangered plant species. A total of 2,407 plant species from 1,171 genera of 338 families have been monitored. Among them, there are 55 species under state priority protection (including introduced plants), such as *Taxus wallichiana var. mairei*, *Bretschneidera sinensis*, and *Alsophila spinulosa*; 83 rare and endangered species, including *Houpoea officinalis* and *Pteroceltis tatarinowii*; and 36 model plants, such as *Cornus hongkongensis subsp. melanotricha*, *Scutellaria tsinyunensis*, and *Acer lucidum*.

There are 1,071 animal species, including 207 species of vertebrates and 864 species of invertebrates. Notably, 18 wild animal species are under state priority conservation, such as *Prionailurus bengalensis* and *Otus lettia*.

Scenery of Jinyunshan (Viewing Platform)

Garrulax sannio; *Lanius schach*; *Aegithalos concinnus*; *Egretta garzetta*; Landscape around Jinyunshan Station

Aralia elata; *Albizia kalkora*; *Cunninghamia lanceolate*; *Pinus massoniana var. massoniana*

Biodiversity Information

Chrysolophus Pictus

Chrysolophus pictus, a rare and endangered bird species, belongs to the genus *Chrysolophus* in the family *Phasianidae* of the order *Galliformes*. It is under Level-II state priority conservation. Males measure about 1 m in length, while females are approximately 60 cm long. Males are characterized by their golden-yellow plumage and a beautiful, golden-yellow filamentous crest on the head. These birds are usually seen alone or in small groups and prefer habitats with low trees on the hillsides, as well as secondary subtropical broad-leaved and deciduous broad-leaved forests. They inhabit rocky hillsides at elevations ranging from 600 to 1,800 m, frequenting bamboo thickets. This species is distributed in Qinghai, Gansu, Shaanxi, and other parts of China.

Discovery of New Species

Tonsilla Jinyunensis

Tonsilla jinyunensis belongs to the genus *Tonsilla* in the family *Agelenidae* and was discovered in Jinyunshan in 2022.

The male spider has a body length of 15.65 cm. Its cephalothorax measures 8.27 cm in length and 4.92 cm in width, while its abdomen is 7.01 cm long and 4.07 cm wide. The carapace is pear-shaped, brown in color, with distinct neck and radial grooves, and the fovea is oriented longitudinally. The female spider has a body length of 16.05 cm. Its cephalothorax measures 8.25 cm in length and 4.86 cm in width, while its abdomen is 8.28 cm long and 5.66 cm wide. It does not have genital structures (epigyne).

Leptobrachella Jinyunensis

Leptobrachella jinyunensis belongs to the genus *Leptobrachella* in the family *Megophryidae*, and is a small frog species that inhabits areas near forested mountain streams with good water quality. It is nocturnal in nature. The new species has the following characteristics. The male measures 29.1–34.1 mm in length and the female 34.1–34.9 mm. The back skin is rough with warts similar in size to the shoulder glands, without conical spines. The ventral side is primarily floral white, covered with deep gray fine spots. The dorsum is dark gray-brown with smoke-black patterns. When the hind limbs are extended forward along the body, they lie closely against it. The frequency at which they sing falls within the range of 4.08 ± 0.16 Hz. This species is mainly distributed at elevations of 343 to 937 m in Jinyunshan.

Chapter 18
Puding Karst Comprehensive Ecological Quality Monitoring Station, Guizhou Province (Forest)

Department of Ecology and Environment Monitoring, Ministry of Ecology and Environment of the People's Republic of China and China National Environmental Monitoring Centre, *The Eye on Ecology: China's Ecological Monitoring Records (Volume 1)*, https://doi.org/10.1007/978-981-95-0047-5_18

Location of the Station

Shawan Group, Longga Village, Chengguan Town, Puding County, Anshun City, Guizhou Province.

Lead Organization

Institute of Geochemistry, Chinese Academy of Sciences.

Participating Organizations

Ecological and Environmental Monitoring Center of Guizhou Province.
Chinese Research Academy of Environmental Sciences.

Key Monitoring Area

Other key ecosystem service zone.

Latitude and Longitude

105° 45′ E.

26° 22′ N.

Representativeness and Typicality

- Center of Southwest China Karst and a typical fragile karst ecosystem in Southwest China.
- Representatives of karst soil conservation functional areas in central Guizhou and karst soil conservation functional areas in Guizhou and Guangxi.
- Representatives of ecological shields and water source conservation areas of the Yangtze River and the Pearl River.
- Typical plateau-centered ecosystem that encompasses multiple types of karst landscapes.

Puding Karst Comprehensive Ecological Quality Monitoring Station ("Puding Karst Station" for short) is located in the center of Southwest China Karst and the transition zone between the Yangtze River and Pearl River systems, representing the typical karst landforms of South China. Puding Karst Station focuses on monitoring the karst ecosystems in South China. It conducts long-term, fixed-point monitoring of material cycling and biogeochemical processes driven by energy flows within these ecosystems. The station examines the structure of primary functional communities and indicative biological communities in various habitats, such as forests, grasslands, urban and rural areas, and farmlands. It also assesses water ecological conditions including water resources, aquatic organisms, aquatic habitats, and aquatic environments. Furthermore, the station studies forest, soil, hydrology, water quality, air pollution, and carbon flux in the region, as well as the responses and feedback mechanisms to climate change and land use. It explores the stability, sustainable management, and evolutionary trends of karst ecosystems against the background of global change and human activity, and serves as an international comparative cooperation research platform and a science education base for ecosystem research in the karst regions of southwestern China and similar environmental areas worldwide.

Building of Puding Karst Station

An aerial view of Puding Karst Station

Chenqi small watershed observation site

Tianlongshan biological monitoring plot

Key karst zone water-carbon-nitrogen flux simulation, observation, and test site and science outreach plank road

Scenery of the Yelang Lake

Biodiversity Information

Ecological restoration in karst areas leads to an increase in biodiversity. Thanks to rocky desertification control efforts, the biodiversity in karst areas has significantly improved. The habitats within these areas harbor a rich variety of life, and the activity levels of wildlife indirectly indicate the effectiveness of the rocky desertification control efforts. The region where the monitoring station is located represents a degraded secondary forest ecosystem in the karst mountainous area of the Yunnan-Guizhou Plateau. The vegetation under the flux tower is mainly composed of degraded vine and thorn shrublands, with patches of secondary evergreen and deciduous broad-leaved forests. The dominant species include *Cunninghamia lanceolata*, followed by pine, catalpa, toona, and poplar.

Danaus Genutia Cramer

Danaus genutia Cramer is one of the dominant butterfly species found in karst areas. It is a medium to large-sized butterfly belonging to the genus *Danaus* in the family *Nymphalidae*. The base color of its wings is orange-yellow or orange-red, while the leading edge, outer half, trailing edge, and wing veins are dark brown. The typically orange or yellow wings are adorned with black stripes and spots. *Danaus genutia Cramer* flies quickly and prefers ample sunlight. It is distributed in Henan, Xizang, Jiangxi, Zhejiang, Fujian, Sichuan, Yunnan, Guangxi, Taiwan, Guangdong, and Hainan. Adult butterflies are often seen flying in forest clearings and around flowers, feeding on nectar. The larvae feed on specific plants and serve as important pollinators in the ecosystem. This butterfly is sensitive to environmental changes and acts as an indicator species for environmental quality. The changes in its population indicate the health status of the eco-environment.

1. Before the battle against rocky desertification.
2. After the battle against rocky desertification.

Prionailurus Bengalensis

Prionailurus bengalensis is a guardian of karst ecosystems and is a small felid species included in China's List of Wild Animals Under State Priority Conservation (Level II) and CITES Appendix II. It is widely distributed in forests, grasslands, and wetlands in Asia. After years of rocky desertification control and ecological restoration efforts in the karst regions of Southwest China, the habitat of *Prionailurus bengalensis* has gradually improved, and this charming and enigmatic creature has made a comeback.

Broussonetia Papyrifera

Broussonetia papyrifera belongs to the genus *Broussonetia* in the family *Moraceae*. It is a deciduous tree or shrub that can reach heights of up to 15 m. It is widely distributed across Southern China and Southeast Asia, such as Yunnan, Guangdong, Guangxi, and Fujian. Known for its high adaptability, it can grow in a variety of soil conditions. The leaves are large and heart-shaped. The fruits mature in summer and autumn, forming umbellate cymes. The aggregate fruits turn orange when ripe, adding ornamental value to the plant.

Broussonetia papyrifera holds significant ecological and economic importance. Its bark is rich in cellulose, which makes it ideal for paper production, hence the nickname "paper mulberry tree." The light and soft wood is a suitable furniture and construction material. Furthermore, the tree's extensive root system aids in preventing soil erosion, playing a vital role in soil conservation and ecological restoration.

Chapter 19
Baima Snow Mountain Comprehensive Ecological Quality Monitoring Station, Yunnan Province (Forest)

Department of Ecology and Environment Monitoring, Ministry of Ecology and Environment of the People's Republic of China and China National Environmental Monitoring Centre, *The Eye on Ecology: China's Ecological Monitoring Records (Volume 1)*, https://doi.org/10.1007/978-981-95-0047-5_19

Location of the Station

Deqin County, Diqing Tibetan Autonomous Prefecture, Yunnan Province.

Quzonggong, Baima Snow Mountain National Nature Reserve.

Lead Organization

Ecological and Environmental Monitoring Center of Yunnan Province.

Participating Organizations

Quzonggong Ecological Positioning Observation and Research Station, Baima Snow Mountain National Nature Reserve.

Diqing Prefecture Ecological and Environmental Monitoring Station, Department of Ecology and Environment of Yunnan Province.

Key Monitoring Area

Northwest Yunnan Plateau Biodiversity Conservation and Water Source Conservation Functional Area.

Latitude and Longitude

99° 2′ E.

28° 19′ N.

Representativeness and Typicality

- One of the world's most biodiverse areas for alpine plants.
- Complete vertical vegetation spectrum.
- Typical alpine and subalpine forest ecosystem.
- Key water source conservation area in the upper reaches of the Yangtze River.

Baima Snow Mountain Comprehensive Ecological Quality Monitoring Station ("Baima Snow Mountain Station" for short) is located in Quzonggong of Baima Snow Mountain National Nature Reserve, the source of the Zhubaluo River, a primary tributary of the Jinsha River. The monitoring station is situated at an elevation of 3,908 m and in a distinctive vertical climate zone, with a complete vertical vegetation spectrum.

The station is equipped with monitoring plots for the vertical vegetation spectrum and a network of infrared cameras for animal monitoring. It also possesses experimental and in-situ monitoring equipment that support satellite remote sensing, low-altitude drone monitoring, and ground-based monitoring, and is capable of air-space-land integrated ecological monitoring.

This region has a high biodiversity, hosting more than 65 percent of China's *Rhinopithecus bieti* population. There are 22 animal species under Level I state priority conservation, including 11 species of mammals such as *Rhinopithecus bieti, Profelis temmincki, Neofelis nebulosa, Panthera pardus, Uncia uncia*, and 11 bird species like *Haliaeetus albicilla, Lophophoruslhuysii*, and *Syrmaticus humiae.* There are two wild plant species under Level-I state priority conservation, namely, *Davidia involucrata var. vilmoriniana* and *Taxus florinii.*

Alpine shrub-meadow on the scree slope of Baima Snow Mountain

Rhodiola crenulata; *Saussurea laniceps*; *Saussurea medusa*; *Meconopsis rudi*

Ciconia nigra; *Gypaetus barbatus*; *Moschus berezovskii*; *Panthera pardus*; *Tetraosis szechenyii*

Biodiversity Information

Rhinopithecus Bieti

Rhinopithecus bieti, which belongs to the genus *Rhinopithecus*, is a rare and endangered primate endemic to China and under Level-I state priority conservation. It is the highest-living primate in the world and found only in a narrow region of the Yunling Mountains, located between the Jinsha River and the Lancang River in the southeastern part of the Qinghai-Xizang Plateau and the central area of the Hengduan Mountains. Monitoring data in recent years shows that there are a total of 23 *Rhinopithecus bieti* communities across the country, with population numbers ranging from 3,360 to 4,330. Specifically, Baima Snow Mountain National Nature Reserve and its surrounding areas are home to 14 communities with a size between 2,180 and 2,830, accounting for more than 65 percent of the total population.

Patrol site of Baima Snow Mountain Station

Chapter 20
Xishuangbanna Comprehensive Ecological Quality Monitoring Station, Yunnan Province (Forest)

Department of Ecology and Environment Monitoring, Ministry of Ecology and Environment of the People's Republic of China and China National Environmental Monitoring Centre, *The Eye on Ecology: China's Ecological Monitoring Records (Volume 1)*, https://doi.org/10.1007/978-981-95-0047-5_20

Location of the Station

East Zone of Xishuangbanna Tropical Botanical Garden, Chinese Academy of Sciences, Menglun Town, Mengla County, Xishuangbanna Dai Autonomous Prefecture, Yunnan Province.

Lead Organization

Xishuangbanna Tropical Botanical Garden, Chinese Academy of Sciences.

Key Monitoring Areas

South Yunnan Biodiversity Conservation Functional Area.

Sichuan-Yunnan Forest and Biodiversity Conservation Functional Area.

Latitude and Longitude

101° 16.4′ E.

21° 55.8′ N.

Representativeness and Typicality

- Typical tropical rainforest and monsoon rainforest ecosystem.
- Transition zone from paleotropical flora to pan-Arctic flora.
- Transition Zone from East Asian flora to Himalayan flora.
- Globally recognized biodiversity hotspot.

Xishuangbanna Comprehensive Ecological Quality Monitoring Station ("Xishuangbanna Station" for short) is located on Hulu Island, Menglun Town, Mengla County, Xishuangbanna Dai Autonomous Prefecture, Yunnan Province (within Xishuangbanna Tropical Botanical Garden, Chinese Academy of Sciences). Xishuangbanna Station consists of the Menglun main station and the Bubang working station. The station focuses on frontier questions in tropical forest ecology, and has established a comprehensive field observation and research platform, an indoor analysis and sample preservation platform, and an information collection, sharing, and management platform, all of which are in stable operation. The station also has 10 online long-term observation plots, one meteorological observation field, and 10 surface runoff fields. The Bubang working station has 20 hectares of dynamic monitoring plots for tropical seasonal rainforests, an 81-m-high canopy crane, and three forest observation towers. It conducts comprehensive monitoring of biodiversity in plants and animals and performs observations of carbon and water fluxes within the ecosystem.

A distant view of the Menglun main station

Rubber forest

Ceyx erithacus; *Rhacophorus kio*

Sapria himalayana; *Alsophila spinulosa*; *Antiaris toxicaria*

Biodiversity Information

Parashorea Chinensis

Parashorea chinensis is a rare and endangered plant under Level-I state priority conservation in China. It is one of the iconic species of the tropical rainforest in Xishuangbanna. Mature trees can reach heights of about 60 m, with the first branches generally starting at heights of 20 m or above. The flowering period lasts from April to June, followed by a fruiting period from July to September. The wood is relatively dense, making it an excellent choice for timber. This species is primarily found in the moist valleys and slopes of southern Mengla County, where it dominates the canopy layer, forming a monodominant community.

Elephas Maximus

Elephas maximus is the largest wild proboscis animal in Asia and a wild animal species under Level-I state priority conservation in China. In 1986, it was classified as an Endangered (EN) species on the IUCN Red List of Threatened Species. According to the report presented at the Third Asian Elephant Range States Meeting held in Nepal in 2022, the global population of *Elephas maximus* is estimated to be between 45,522 and 48,671. Countries such as Laos, China, and Nepal each have fewer than 400 each. In China, wild populations of *Elephas maximus* are found exclusively in Xishuangbanna Dai Autonomous Prefecture, Pu'er City, and Lincang City in Yunnan.

Elephas maximus

Chapter 21
Nyingchi Comprehensive Ecological Quality Monitoring Station, Xizang (Forest)

Department of Ecology and Environment Monitoring, Ministry of Ecology and Environment of the People's Republic of China and China National Environmental Monitoring Centre, *The Eye on Ecology: China's Ecological Monitoring Records (Volume 1)*, https://doi.org/10.1007/978-981-95-0047-5_21

Location of the Station

Jiangnan Road, Bayi District, Nyingchi City.

Lead Organization

Xizang Center of Remote Sensing for Ecological and Environmental Monitoring.

Participating Organizations

Xizang Agricultural and Animal Husbandry University.

Chinese Research Academy of Environmental Sciences.

Institute of Tibetan Plateau Research, Chinese Academy of Sciences.

Key Monitoring Area

Southeast Xizang Biodiversity Conservation Key Functional Area.

Latitude and Longitude

94° 22′ E.

29° 35′ N.

Representativeness and Typicality

- The most complete vertical vegetation spectrum in the world.
- One of the world's Key Biodiversity Areas.
- A vital source for the global ecosystem and an initiator and regulator of climate change.

Nyingchi Comprehensive Ecological Quality Monitoring Station ("Nyingchi Station" for short) is located in the Bayi District of Nyingchi City in Xizang. The station consists of multiple monitoring sites. The unique geographical environment has given rise to a rich ecosystem, making it one of the world's Key Biodiversity Areas. It boasts the most complete vertical vegetation spectrum in the world. Nyingchi Station is dedicated to studying the patterns of vegetation succession and the ways organisms respond and adapt to environmental changes in the context of global climate change. It is equipped with comprehensive infrastructure that supports integrated environmental monitoring, including flux observation systems, a weather observation station, soil moisture monitoring systems, soil profile CO_2 observation systems, bird observation systems, wildlife tracking systems, automated insect monitoring systems, as well as multiple portable analyzers. These facilities enable a holistic approach to monitoring water, soil, air, and biological components.

Alpine meadow

Zhagonggou Weather Observation Site

Pseudois nayaur; *Saxicola stejnegeri*; *Carduelis ambigua*; *Eumyias thalassinus*

Accipiter nisus; *Falco tinnunculus*; *Arborophila torqueola*; *Psittacula derbiana*; *Hedychium yunnanense*; *Aconitum hemsleyanum*; *Incarvillea mairei*; *Campanula pallida*

Biodiversity Information

Cupressus Gigantea

Cupressus gigantea, which is endemic to Xizang, belongs to the genus *Cupressus* in the family *Cupressaceae*. This species is under Level-I state priority conservation in China and classified as a Vulnerable (VU) species on the IUCN Red List of Threatened Species. The bark is longitudinally fissured into strips; the branches with scaly leaves are closely arranged, thick, quadrangular, often covered with wax powder, and the terminal scaly leaves do not droop. There are narrow wings on both sides of the seed. This species has an extremely small population, found only in Nangxian County, Mainling County, and other parts of Nyingchi City in the Yarlung Zangbo River Basin. The trees typically grow at elevations between 3,000 and 3,400 m, and form sparse, pure stands in floodplains along the river and the middle to lower sections of south-facing terraces with exposed limestone outcrops. In the *Cupressus gigantea* Nature Reserve in Bajie Town of Nyingchi City, there stands a cypress tree whose trunk is wide enough to require a dozen people to lock hands to circle it. It towers over 50 m in height, with a diameter of nearly 6 m, and its crown casts a shadow covering more than 666.67 square meters. Calculations suggest that this cypress tree is between 2,000 and 2,500 years old. Legend has it that the tree was planted by Songtsen Gampo and Princess Wencheng of the Tang Dynasty (618–907) on their wedding day and is revered as a "sacred tree" by the locals.

Grus Nigricollis

Grus nigricollis, known as the black-necked crane, is a large wading bird belonging to the genus *Grus* in the family *Gruidae*, and is under Level-I state priority conservation in China. The bird gets its name as the majority of its head and the upper two-thirds of its neck is black, with the exception of a small white or grayish-white patch behind and below the eyes. This species measures around 120 cm in body length and is distinguished by its long beak, neck, and legs, and tall stature. *Grus nigricollis* lives in wetlands such as plateau marshes, lakes, and river beaches at elevations ranging from 2,500 to 5,000 m. It is classified as a Near Threatened (NT) species on the IUCN Red List of Threatened Species and also included in CITES.

Chapter 22
Qinling Comprehensive Ecological Quality Monitoring Station, Shaanxi Province (Forest)

Department of Ecology and Environment Monitoring, Ministry of Ecology and Environment of the People's Republic of China and China National Environmental Monitoring Centre, *The Eye on Ecology: China's Ecological Monitoring Records (Volume 1)*, https://doi.org/10.1007/978-981-95-0047-5_22

Location of the Station

Jixian Town, Zhouzhi County, Xi'an City, Shaanxi Province.

Lead Organization

Shaanxi Academy of Sciences.

Participating Organizations

Qinling National Botanical Garden.

Environmental Monitoring Center of Shaanxi Province.

Shaanxi University of Science and Technology.

Key Monitoring Areas

Qinling Mountains Water Source Conservation and Biodiversity Conservation Functional Area.

Daba Mountains Biodiversity Conservation Functional Area.

Latitude and Longitude

108° 20′ E.

34.3° 19′ N.

Representativeness and Typicality

- An important water source for the South-to-North Water Diversion Project and a "Central Water Tower" of China.
- Globally recognized biodiversity hotspot.
- Geographical and cultural boundary between North and South China.

Qinling Comprehensive Ecological Quality Monitoring Station ("Qinling Station" for short) lies within the Qinling National Botanical Garden in Zhouzhi County, Xi'an City. With a planned area of 639 square kilometers, the region where the station is located covers five geomorphic types: high mountains, medium mountains, low mountains, hills, and plains. The elevation difference is 2,417 m, leading to significant vertical climate variation. This region is characterized by extremely rich biodiversity and features a complete forest ecosystem along with remnants of Quaternary glaciers.

The station has complete monitoring facilities, with nearly 1,000 square meters of laboratory and office space. It is equipped with over 40 types of monitoring devices and instruments, such as infrared-triggered cameras, satellite tracking collars, and LiDAR drones for biodiversity monitoring. The station also has online water quality monitoring devices, fully automated discrete chemical analyzers, inductively coupled plasma optical emission spectroscopy (ICP-OES), and gas chromatography-mass spectrometry (GC-MS) systems for water quality analysis. Over the past 60 years, the station has amassed massive amounts of basic data on the wildlife and natural resources of the Qinling mountain range.

Landscape around Qinling Station

Vegetation in Qinling National Botanical Garden; Sample plot of Qinling Station

Tragopan temminckii; *Nipponia nippon*; *Chrysolophus pictus*; *Ithaginis cruentus*

Rhinopithecus roxellana; *Paeonia rockii subsp. atava*; *Moschus berezovskii*; *Budorcas bedfordi*; *Davidia involucrata*; *Taxus wallichiana var.chinensis*; *Psathyrostachyshuashanica*

Biodiversity Information

Ailuropoda Melanoleuca

Ailuropoda melanoleuca, commonly known as the giant panda, is a rare relic species endemic to China and referred to as China's "national treasure." Known for its striking black-and-white coat and bear-like appearance, this ancient species first emerged on the southern slopes of the Qinling Mountains during the early Pleistocene epoch. By the middle of the Pleistocene, it had expanded its range northward beyond the Qinling mountain range. Among the members of the *Ailuropoda-stegodon* assemblage, only the giant panda has survived to the present day, hence hailed as a "living fossil." *Ailuropoda melanoleuca* is endemic to Sichuan, Gansu, and Shaanxi in China. In Shaanxi Province, they mainly live in Foping, Yangxian, Taibai, Zhouzhi, and Ningshan counties along the Qinling mountain range. Qizai, the world's only brown-and-white panda in captivity, is considered a "treasure among treasures" due to its rare brown-and-white fur. To date, brown pandas have only been discovered and monitored 10 times, all within the Qinling mountain range.

Chapter 23
Loess Plateau Comprehensive Ecological Quality Monitoring Station, Shaanxi Province (Forest)

Department of Ecology and Environment Monitoring, Ministry of Ecology and Environment of the People's Republic of China and China National Environmental Monitoring Centre, *The Eye on Ecology: China's Ecological Monitoring Records (Volume 1)*, https://doi.org/10.1007/978-981-95-0047-5_23

Location of the Station

Luochuan County, Yan'an City, Shaanxi Province.

Lead Organization

Institute of Earth Environment, Chinese Academy of Sciences.

Participating Organizations

Xi'an Institute for Innovative Earth Environment Research.

Research Center for Eco-Environmental Sciences, Chinese Academy of Sciences.

Key Monitoring Area

Loess Plateau Soil Erosion Control Area.

Latitude and Longitude

109° 26′ E.

35° 43′ N.

Representativeness and Typicality

- Typical Loess Plateau landform.
- Representative biodiversity.
- Deciduous broad-leaved forest of the warm-temperate zone.
- Typical soil and water conservation functional area of the Loess Plateau.

Loess Plateau Comprehensive Ecological Quality Monitoring Station ("Loess Plateau Station" for short) is located in Luochuan County, Yan'an City, Shaanxi Province. Situated in the world's largest continuous loess zone, this region has the highest levels of soil erosion and the most fragile ecosystems. Loess Plateau Station primarily monitors key ecological engineering areas for soil erosion control. It adopts a combination of satellite remote sensing, aerial remote sensing, and ground-based monitoring methods, and is capable of air-space-land integrated ecological monitoring. The station has comprehensive observation fields, observation zones, observation points, surface flux observation sites, groundwater monitoring wells, and other observation facilities. With over 100 sets of monitoring instruments, the station is capable of conducting long-term monitoring of key ecosystem components including water, soil, air, biological elements, and interface processes in the region. From the perspective of Earth system science, the station focuses on critical environmental issues such as regional biodiversity, climate change, the effects of management practices on the ecological quality of planted forests, as well as the improvement of ecosystem structure and functions.

Winter scenery of Xiangsichuan Forest Park in Luochuan County

Biodiversity Information

Lilium Pumilum

Lilium pumilum is a perennial herb belonging to the genus *Lilium* in the family *Liliaceae*. It features a white underground bulb that is ovoid or conical in shape. The leaves are linear and scattered along the middle part of the stem. The flowers are red or purplish-red. This species is commonly found at elevations above 600 m on sunny slopes or in sparse forests and grasslands. *Lilium pumilum* has beautiful, dainty flowers that are highly ornamental. This species is also a source of cold-resistant germplasm in the genus *Lilium*, suitable for breeding cold-hardy and drought-tolerant lily varieties.

Cotoneaster Multiflorus

Cotoneaster multilorus is a deciduous shrub belonging to the genus *Cotoneaster* in the family *Rosaceae.* It features slender, often arching branches, with cylindrical twigs that are reddish-brown or brownish, hairless, and may have short bristles when young, which will soon fall off. The leaves are ovate or broadly ovate, measuring 2–4 cm in length and 1.5–3 cm in width. This species is generally found in mixed forests in valleys and on slopes at elevations between 1,200 and 3,500 m. It is cold-hardy, loves the sun but can survive in partial shade, is extremely drought-tolerant, and can thrive in poor soils. It grows best in fertile, well-drained loam soils with good aeration.

Chapter 24
Qilianshan Comprehensive Ecological Quality Monitoring Station, Gansu Province (Forest)

Department of Ecology and Environment Monitoring, Ministry of Ecology and Environment of the People's Republic of China and China National Environmental Monitoring Centre, *The Eye on Ecology: China's Ecological Monitoring Records (Volume 1)*, https://doi.org/10.1007/978-981-95-0047-5_24

Location of the Station

Sunan Yugur Autonomous County, Zhangye City, Gansu Province.

Qilianshan National Nature Reserve Pilot Area.

Lead Organization

Ecological and Environmental Monitoring Center of Zhangye City.

Participating Organizations

Ecological and Environmental Monitoring Center of Gansu Province.

Northwest Institute of Eco-Environment and Resources, Chinese Academy of Sciences.

Key Monitoring Area

Hexi Corridor Key Functional Area of Wind Prevention and Sand Fixation.

Latitude and Longitude

100° 28′ E.

38° 25′ N.

Representativeness and Typicality

- Key functional area of biodiversity conservation.
- Complete vertical vegetation spectrum.
- Key biological germplasm bank in Northwest China.
- The source of the inland Heihe River, Shule River, and Shiyang River.
- Key recharge area of the Yellow River and the Qinghai Lake.

Qilianshan Comprehensive Ecological Quality Monitoring Station ("Qilianshan Station" for short) is situated in the Qilianshan National Nature Reserve Pilot Area, Sunan Yugur Autonomous County, Zhangye City, Gansu Province. This region is an important ecological security shield in Northwest China. Qilianshan Station covers an area of nearly 4,000 m^2, with a floor space of about 700 m^2. This station is equipped with devices and facilities for monitoring meteorology, greenhouse gas fluxes, radiation, biodiversity, air quality, water quality, soil conditions, etc. With the aid of satellite remote sensing, the station implements comprehensive, multi-dimensional eco-environmental monitoring of the region, enabling field experiments, system simulations, and integrated analysis. The station aims to reveal carbon and water cycling processes and their interactions at various scales ranging from individual plants and plant communities to ecosystems, landscapes, and watersheds, and gain a precise understanding of the intrinsic changes and mechanisms within the regional eco-environment to facilitate refined monitoring, assessment, and early warning of environmental changes.

Qilian Mountains

Przewalskium albirostris; *Vulpes ferrilata*; *Grus nigricollis*; *Pseudois nayaur*; *Crossoptilon auritum*; *Uncia uncia*; *Cervus canadensis*

Rhododendron anthopogonoides; *Stellera chamaejasme*; *Halenia elliptica*; *Caragana jubata*; *Pinus tabuliformis*; *Gentiana siphonantha*; *Iris lactea*; *Iris lactea*; *Rosa omeiensis f. pteracantha*; *Gentiana dahurica*; *Rheum uninerve*

Part II
Grassland Impression

Chapter 25
Bashang Agro-Pastoral Ecotone Station, Hebei Province (Grassland)

Department of Ecology and Environment Monitoring, Ministry of Ecology and Environment of the People's Republic of China and China National Environmental Monitoring Centre, *The Eye on Ecology: China's Ecological Monitoring Records (Volume 1)*, https://doi.org/10.1007/978-981-95-0047-5_25

Location of the Station

Guyuan County and Kangbao County, Zhangjiakou City, Hebei Province

Lead Organization

Ecological and Environmental Monitoring Center of Hebei Province

Participating Organization

Institute of Geographic Sciences, Hebei Academy of Sciences

China Agricultural University

Key Monitoring Area

Bashang Grasslands Key Functional Area of Wind Prevention and Sand Fixation

Latitude and Longitude

Guyuan Station

115° 40′ E

41° 46′ N

Kangbao Station

114° 48′ E

42 ° 7′ N

Weichang Station

116° 58′ E

42° 8′ N

Representativeness and Typicality

- Typical northern agro-pastoral ecotone.
- Ecological security shield for the capital city.
- Key functional area of wind prevention and sand fixation.

Bashang Agro-Pastoral Ecotone Station consists of a monitoring station in Guyuan County and Kangbao County of Zhangjiakou City of Hebei Province, respectively, with a total floor space of about 5,000 m^2 and long-term observation sites. This region has numerous lakes and wetlands and serves as a habitat for many wildlife species under state priority conservation, including *Larus relictus*, *Ciconia nigra*, *Ciconia boyciana*, *Otis tarda*, *Lyrurus tetrix*, and *Grus virgo*. The region includes several nature reserves such as Saihanba National Nature Reserve, Hongsongwa National Nature Reserve, Luanhe River Source Provincial Nature Reserve, Yudaokou Provincial Nature Reserve, Kangbanaoer National Wetland Park, and Huangtuwan National Grassland Park. These nature reserves play a crucial role in biodiversity conservation and provide essential ecological services such as carbon sequestration and oxygen release.

Guyuan Station covers a farmland of 1,000 mu (66.67 ha) and a natural grassland of 200 mu (13.33 ha) and has established monitoring platforms for the restoration of cultivated grassland and degraded grassland. Kangbao Station has established an observation field covering nearly 200 mu (13.33 ha), featuring long-term observation sites for farmland, grassland, and shrub-grassland ecosystems. The station is equipped with a 30-m-high observation tower, a weather station, and over 50 other instruments. These facilities enable long-term monitoring of vegetation, water, soil, and air within the ecosystem. The station also uses satellite remote sensing and drones to monitor ecosystem changes and conduct ground validation of remote sensing data.

Farmland in summer

The pillar

Birding at the Tian'er Lake

Aythya nyroca; *Tringa erythropus*

Pica hudsonia; *Urocissa erythrorhyncha*; *Delphinium grandiflorum*; *Achnatherum splendens*

Biodiversity Information

Larus Relictus

Larus relictus is a medium-sized waterbird belonging to the genus *Larus* in the family *Laridae* of the order *Charadriiformes*. The bird measures 39 to 46 cm in length. Its bill is robust and dark red; its iris is brownish and there is a crescent-shaped white patch above and below each eye, forming a broad white eye ring. The forehead is flat. In its summer plumage, the head is dark brown to black, making a distinctive dark hood. The back and shoulders are pale gray, while the outer primary feathers are white with black subterminal bands. In flight, it can be identified by small patches of white feathers at the tips of the wings. The rump, tail feathers, and underparts are white. The legs are dark red or coral red. In its winter plumage, the head turns white, with only a small dark patch in the ear region and a slightly darker shade on the crown to the nape. Juvenile birds in their first winter plumage share some similarities with adult winter plumage but differ in several distinctive features: The ear coverts do not have the dark patch seen in adults, there is a dark crescent-shaped mark in front of the eye, the nape has dark longitudinal streaks, the tertiary feathers and some wing coverts are dark brown, and the tail feathers are white with a broad black terminal band at the tips. It is under Level-I state priority conservation and classified as a Vulnerable species.

Chapter 26
Xilingol Comprehensive Ecological Quality Monitoring, Inner Mongolia (Grassland)

Department of Ecology and Environment Monitoring, Ministry of Ecology and Environment of the People's Republic of China and China National Environmental Monitoring Centre, *The Eye on Ecology: China's Ecological Monitoring Records (Volume 1)*, https://doi.org/10.1007/978-981-95-0047-5_26

Location of the Station:

Xilinhot City, Xilin Gol League, Inner Mongolia Autonomous Region.

Lead Organization:

Environmental Monitoring Center of Inner Mongolia.

Participating Organizations:

Institute of Botany, Chinese Academy of Sciences.

Inner Mongolia University.

Key Monitoring Area:

Hunshandake Desertification Control Area.

Latitude and Longitude:

116° 42′ E.

43° 3.9′ N.

Representativeness and Typicality

- Typical temperate grassland ecosystem
- Ecological security shield in northern China
- Key functional area of wind prevention and sand fixation.

Established in 1979, Xilin Gol Comprehensive Ecological Quality Monitoring ("Xilingol Station" for short) is the first long-term ecosystem research station in the temperate grassland region of China, located in Xilin Gol League, central Inner Mongolia. This region is the most typical and representative region for grassland ecosystems in northern China.

This station focuses on studying the structure and function of grassland ecosystems, the restoration of degraded grasslands, etc. It boasts comprehensive infrastructure, including over a dozen observation sites and experiment and research platforms, as well as sample plots with a total area of more than 650 hectares. This region hosts 856 species of higher plants belonging to 386 genera of 116 families. They include 713 species of seed plants in 314 genera of 80 families, 115 species of bryophytes in 58 genera of 26 families, and 28 species of ferns in 14 genera of 10 families. There are 671 species of herbaceous plants and 42 species of woody plants, accounting for 94.11 percent and 5.89% of the total species in the flora, respectively. Additionally, there are 228 species of vertebrates belonging to 62 families of 27 orders.

Leymus chinensis grassland

Poplar and birch forest in Xilin Gol Grassland National Nature Reserve

Limonium bicolor demonstration area; *Phragmites australis*; *Ruta graveolens*; *Oxytropis racemosa*; *Spiranthes sinensis*; *Scabiosa tschiliensis*; *Echinops sphacroccphalus*; *Linaria vulgaris subsp. chinensis*

Saposhnikovia divaricata; *Artemisia frigida*; *Pseudolysimachion incanum*; *Thymus mongolicus*; *Orostachys fimbriata*; *Stellera chamaejasme*; *Delphinium grandiflorum*; *Ligularia sibirica*; *Lilium pumilum*; *Adenophora stenanthina*

Chapter 27
Hulun Buir Comprehensive Ecological Quality Monitoring Station, Inner Mongolia (Grassland)

Department of Ecology and Environment Monitoring, Ministry of Ecology and Environment of the People's Republic of China and China National Environmental Monitoring Centre, *The Eye on Ecology: China's Ecological Monitoring Records (Volume 1)*, https://doi.org/10.1007/978-981-95-0047-5_27

Location of the Station

Hag Town, Hailar District, Hulun Buir City, Inner Mongolia Autonomous Region.

Lead Organization

Environmental Monitoring Center of Inner Mongolia.

Participating Organizations

Institute of Agricultural Resources and Regional Planning (IARRP), Chinese Academy of Agricultural Sciences.

Chinese Research Academy of Environmental Sciences.

Inner Mongolia Huihe National Nature Reserve Administration.

Key Monitoring Area

Hulun Buir Meadow Steppe Conservation Area.

Latitude and Longitude

119° 55′ E.

49° 19′ N.

Representativeness and Typicality

- Typical meadow steppe ecosystem
- Key functional area of wind prevention and sand fixation
- Key functional area of biodiversity conservation.

Hulun Buir Comprehensive Ecological Quality Monitoring Station ("Hulun Buir Station" for short) is located in Hag Town, Hailar District, Hulun Buir City, Inner Mongolia Autonomous Region. Focusing on studying temperate meadow steppes, this station belongs to a region that has superior hydrothermal conditions among China's grassland ecosystems and features high productivity and rich biodiversity. The primary vegetation types include *Leymus chinensis*, *Stipa baicalensis*, and *Filifolium sibiricum* meadow steppes. The station has established several long-term experiment platforms up to international standards, including those for long-term grazing, long-term mowing, grassland improvement, cultivated grassland, and large-scale ecological remote sensing test sites. The station has gathered more than 30 million entries of time-series monitoring data, over 35,000 soil and plant samples, and more than 3,000 plant specimens.

Grazing platform of Hulun Buir Station

Hulun Buir Station

Stipa baicalensis meadow steppe.

Biodiversity Information

Leymus Chinensis

Leymus chinensis is a species of bunchgrass in the genus *Leymus* in the family *Gramineae*. Known for its resilience to trampling and grazing, it is particularly favored by sheep and goats, hence its common name false wheatgrass. It is generally distributed in cold regions of Northeast China and Russia and now widely cultivated in Northeast China, Northwest China, North China, and Inner Mongolia. *Leymus chinensis* is cold-hardy, tolerant of drought and alkalinity, and capable of growing in poor soils. It is resistant to wind and sand, though it does not tolerate waterlogging. It also exhibits strong adaptability, thriving in plains, on slopes, and in sandy loam soils. This species has plenty of highly palatable leaves with high nutritional value, making it a favorite food for various farm animals all year round. It is an excellent pasture species.

Filifolium Sibiricum

Filifolium sibiricum is a herbaceous perennial belonging to the genus *Filifolium* in the family *Asteraceae*. It grows in grassy mountain slopes and prefers moist conditions. It is cold-hardy and serves as an important constructive species in montane grasslands. *Filifolium sibiricum* forms a widely distributed dominant plant community in the forest-steppe zone and thrives on the upper slopes and summits of low mountains and hills, while it is generally found at higher elevations in typical steppe zones.

Stipa Baicalensis

Stipa baicalensis, commonly known as wolf needle grass, is a perennial dense tussock grass in the genus *Stipa* in the family *Poaceae*. It is a mid-drought-tolerant species typical of meadow steppes, known for its cold resistance and drought tolerance. This grass thrives in well-drained habitats but is not tolerant of saline-alkali conditions, so it is never found on salt-affected soils. *Stipa baicalensis* acts as the dominant species in its own steppe ecosystem, and also commonly appears as a sub-dominant species in other grassland types, such as *Filifolium sibiricum* or *Leymus chinensis* steppes.

Chapter 28
Nagqu Comprehensive Ecological Quality Monitoring Station, Xizang (Grassland)

Experiment facility and Baingoin Grassland

Department of Ecology and Environment Monitoring, Ministry of Ecology and Environment of the People's Republic of China and China National Environmental Monitoring Centre, *The Eye on Ecology: China's Ecological Monitoring Records (Volume 1)*, https://doi.org/10.1007/978-981-95-0047-5_28

Location of the Station

Quguorenmao Village, Nagqu Town, Seni District, Nagqu City in Xizang.

Lead Organization

Xizang Center of Remote Sensing for Ecological and Environmental Monitoring.

Participating Organizations

Institute of Geographic Sciences and Natural Resources Research, Chinese Academy of Sciences.

Satellite Application Center for Ecology and Environment, Ministry of Ecology and Environment.

Key Monitoring Area

Qiangtang Plateau Desertification Control Area in Northwest Xizang.

Latitude and Longitude

92°0′ E.

31° 38′ N.

Representativeness and Typicality

- Alpine grassland ecosystem
- Representative biodiversity
- Key biodiversity conservation area.

Nagqu Comprehensive Ecological Quality Monitoring Station in Xizang ("Nagqu Station" for short) is located on the southern slope of the Tanggula Mountains and the northern foothills of the Nyainqêntanglha Mountains, with an average elevation exceeding 4,500 m. It falls within the subfrigid zone and lies in a transition zone between seasonally frozen ground and permafrost, featuring a typical alpine meadow ecosystem. As an important part of the ecological security shield on the Qinghai-Xizang Plateau, it takes the alpine grasslands of northern Xizang as a model system and pioneering system to research on how biodiversity, community structure, and key functional processes in grassland ecosystems respond to and adapt to climate change and human activity. A series of long-term observation and research platforms have been established, including the thermal gradient platform, atmospheric CO_2 enrichment platform, online nutrient and water replenishment platform, permafrost research platform, and alpine grassland transects.

Starry sky over the experimental site

Biodiversity Information

Cordyceps Sinensis

Cordyceps sinensis is native to Nagqu, Xizang and is a type of Tibetan herbal medicine, or the broader yartsa gunbu, distributed in grasslands. The harvesting period is from May to June. Yartsa gunbu, literally meaning "summer grass, winter worm," is a combination of the larvae of the moth *Hepialus armoricanus* and the *Cordyceps* fungus. Every midsummer, the female moth lays eggs on flowers and leaves. These eggs gradually develop into larvae, which then burrow into wet, loose soils to feed on the nutrients from plant roots. Over time, the larvae grow plump and white, and the spherical ascospores of the *Cordyceps* fungus penetrate the body of the larvae, absorbing nutrients and initiating mycelium growth. As the fungal infection spreads, the infected larvae move up to within 2–3 cm of the soil surface, where they die with their heads pointing upward and tails downward. This is what is known as the "winter worm." Although the larva dies, the fungus inside their body continues to grow, eventually filling the entire insect. In the late spring and early summer of the following year, a purple-red stalk emerges from the head of the dead larva, forming what is known as the "summer grass." From March to May, a stroma* grows from the head of the larva, protruding above the ground. This forms "winter worm, summer grass."

Carex Parvula

Carex parvula is an herbaceous perennial that grows in a cushion shape. The stems are 1–3.5 cm tall, cylindrical and hairless, and have fine ridges, with dense, persistent, brown leaf sheaths at the base. The linear leaves are nearly equal in length to the stems and measure about 0.5 mm wide; they are rigid, with a groove on the ventral side and rough edges. *Carex parvula* primarily grows in alpine meadows, where the soil is mainly alpine meadow soil. The alpine meadow ecosystem dominated by *Carex parvula* features low-growing grasses with a relatively high canopy cover of about 95%. The dense canopy is evenly distributed with no distinct hierarchy, forming a single layer. The turf layer it creates is a continuous mat-like sod layer, about 20 cm thick. This layer is essential for key ecological functions such as soil productivity, water retention, carbon sequestration, and climate regulation in alpine meadows. It plays a crucial role in maintaining the stability of the plateau meadow ecosystem and ensuring the well-being of herders.

* Stroma: It is a fungal structure which supports the true fruiting structure, and represents a transitional mycelial tissue from the vegetative phase to the reproductive phase.

Cordyceps sinensis

Carex parvula.

Chapter 29
Shenzha Comprehensive Ecological Quality Monitoring Station, Xizang (Grassland)

Department of Ecology and Environment Monitoring, Ministry of Ecology and Environment of the People's Republic of China and China National Environmental Monitoring Centre, *The Eye on Ecology: China's Ecological Monitoring Records (Volume 1)*, https://doi.org/10.1007/978-981-95-0047-5_29

Location of the Station

Xainza County, Nagqu City in Xizang.

Lead Organization

Chinese Academy of Sciences.

Institute of Mountain Hazards and Environment (IMHE), Chinese Academy of Sciences, Chengdu.

Participating Organization

Xizang Center of Remote Sensing for Ecological and Environmental Monitoring.

Key Monitoring Area

Qiangtang Plateau Desertification Control Area in Northwest Xizang.

Latitude and Longitude

88° 42′ E.

30° 57′ N.

Representativeness and Typicality

- The world's typical high-elevation comprehensive ecological quality monitoring station and unique observation station of alpine grassland and alpine wetland ecosystems, with strong representativeness
- Ecological security shields in the northern Xizang Plateau and western Xizang mountains
- The ecological security shield projects mainly include the natural grassland protection projects and the wildlife conservation and protected area construction projects.

Wetland

Shenzha Comprehensive Ecological Quality Monitoring Station ("Shenzha Station" for short) is located in Shenzha County, Nagqu City, at an elevation of 4, 675 m. This station is located in the center of the plateau, at the junction of central Xizang, the Gangdise Range, and the Selincuo Lake, the largest lake in northern Xizang, and lies in the southern part of the Qiangtang Plateau, within the alpine grassland-basin region. The station focuses on studying the structure, processes, and functions of alpine grassland and wetland ecosystems. It conducts comprehensive observations of these ecosystems, and is equipped with various monitoring instruments such as a near-surface flux observation station, wind erosion observation system, standard weather station, precipitation form observation system, greenhouse gas flux observation system, soil carbon, temperature, and moisture profile system, and plant spectral analysis system.

Xainza alpine wetland ecosystem observation site

Xainza alpine grassland ecosystem observation site

Biodiversity Information

Stipa Purpurea

Stipa purpurea is a perennial, dense tussock grass belonging to the genus *Stipa* in the family *Poaceae*. It has fine but tough fibrous roots, slender stems, and leaves that are longitudinally rolled like a needle and slightly rough on the underside. The base of its panicle inflorescence is generally enclosed within the leaf sheaths. The fruit is a caryopsis (grain). The flowering and fruiting period occurs from July to October. *Stipa purpurea* is a dominant species in the alpine grassland of Xainza County. The plant has soft, highly palatable stems and leaves. During its flowering period, it is rich in crude protein, fat, and fiber. This grazing-tolerance species is a favorite food for various farm animals all year round, and an excellent forage grass in alpine pastures, while also playing an important role in soil and water conservation and the prevention and control of desertification.

Chapter 30
Qilianshan Comprehensive Ecological Quality Monitoring Station, Qinghai Province (Grassland)

Department of Ecology and Environment Monitoring, Ministry of Ecology and Environment of the People's Republic of China and China National Environmental Monitoring Centre, *The Eye on Ecology: China's Ecological Monitoring Records (Volume 1)*, https://doi.org/10.1007/978-981-95-0047-5_30

Location of the Station

No. 1 Fengxiakou Road, Stud Farm Livestock Station, Qingshizui Town, Menyuan Hui Autonomous County, Haibei Tibetan Autonomous Prefecture, Qinghai Province.

Lead Organization

Northwest Institute of Plateau Biology, Chinese Academy of Sciences.

Key Monitoring Area

Hexi Corridor Key Functional Area of Wind Prevention and Sand Fixation.

Qilianshan (or Qilian Mountains) Glacier and Water Source Conservation Functional Area.

Latitude and Longitude

101° 18′ E.

37° 36′ N.

Representativeness and Typicality

- Alpine *Carex alatauensis* meadow, *Kobresia humilis* meadow, shrub meadow, and swampy meadow ecosystems
- Key ecological security shield and drainage basin
- Priority area of biodiversity conservation
- Representative climate and ecosystem types in Qilianshan.

Qilianshan Comprehensive Ecological Quality Monitoring Station is located on No. 1 Fengxiakou Road, Stud Farm Livestock Station, Qingshizui Town, Menyuan Hui Autonomous County, Haibei Tibetan Autonomous Prefecture, Qinghai Province, at the southern foothills of Qilianshan on the northeastern edge of the Qinghai-Xizang Plateau, at an elevation of 3,250 m. This region has a plateau continental climate, and the vegetation types include alpine shrublands, *Carex alatauensis* meadows, shrub meadows, swampy meadows, and *Kobresia humilis* meadows. It has high plant diversity, hosting 515 species of vascular plants across 210 genera of 56 families. The station primarily conducts research on biodiversity and ecosystem processes and functions, as well as trials on adaptive management of grassland ecosystems and restoration of degraded grasslands. It is equipped with instruments such as an eddy covariance observation system, soil greenhouse gas emission observation system, hydrogen–oxygen isotope analyzer, and gas chromatograph. The station has established a 190-hectare long-term monitoring plot and a carbon–water flux monitoring platform.

Southern foothills of Qilianshan

Summer scenery of Qilianshan

Winter scenery of Qilianshan

Climate change simulation and research platform

Ochotona curzoniae

Pulveroboletus ravenelii; *Meconopsis integrifolia*; *Carex alatauensis*

Linaria flavirostris

Leptopoecile sophiae.

Biodiversity Information

Swertia Przewalskii

Swertia przewalskii is an herbaceous perennial belonging to the genus *Swertia* in the family *Gentianaceae*. It grows to a height of 8–25 cm and has a short rhizome, erect stem, and 1 or 2 pairs of basal leaves, each with a long petiole. The leaves have 3–5 veins and are elliptic, ovate-elliptic to spatulate in shape, with the leaf base narrowing into a petiole. This species grows in shrublands, alpine meadows, swampy meadows, and on damp ground at elevations between 3,000 and 4,200 m.

Flagship Species in the Reserve

Sinadoxa Corydalifolia

Sinadoxa corydalifolia is the only species in the monotypic plant genus *Sinadoxa* in the family *Adoxaceae*, the endemic genus and family to Qinghai Province. It typically grows at elevations between 3,900 and 4,800 m on the lower parts of steep cliffs in seasonal gullies on mountain slopes, in deep crevices of large rocks, or under damp rock walls in alpine scree slopes. It is a typical shade-loving plant with strict habitat requirements, so it has a very narrow distribution range and small wild populations.

Sinadoxa corydalifolia was discovered in 1964 and identified as a new taxonomic group in 1979. Specimens are stored at the Qinghai-Xizang Plateau Biological Specimen Museum of the Northwest Institute of Plateau Biology, Chinese Academy of Sciences. Its discovery upended the 200-plus-year-old belief that there had been only one genus with a single species in the family *Gentianaceae*.

Chapter 31
Ili Valley Comprehensive Ecological Quality Monitoring Station, Xinjiang (Grassland)

Department of Ecology and Environment Monitoring, Ministry of Ecology and Environment of the People's Republic of China and China National Environmental Monitoring Centre, *The Eye on Ecology: China's Ecological Monitoring Records (Volume 1)*, https://doi.org/10.1007/978-981-95-0047-5_31

Location of the Station

Wulasitai Forest Park, Nalati Town, Xinyuan County, Ili Kazakh Autonomous Prefecture, Xinjiang.

Lead Organization

Ecological and Environmental Monitoring Center of Xinjiang.

Participating Organization

Ecological and Environmental Monitoring Center of Ili.

Key Monitoring Area

Tianshan Water Source Conservation and Biodiversity Conservation Key Functional Area.

Latitude and Longitude

84° 12′ E.

43° 17′ N.

Representativeness and Typicality

- Tianshan Water Source Conservation and Biodiversity Conservation Key Functional Area
- Biologically diverse area in the hinterland of the Eurasian continent
- A “wet island” in the arid zone of the Eurasian continent
- Representative mountain-basin vertical ecosystem.

Ili Valley Comprehensive Ecological Quality Monitoring Station (“Ili Valley Station” for short) is located in Nalati Grassland, Xinyuan County, Ili Kazakh Autonomous Prefecture, Xinjiang. This region is home to globally representative temperate alpine and subalpine meadows, mountain meadows, mountain steppes, desert steppes, and other ecosystems, and features a variety of grassland ecosystem types. It is one of the most biologically diverse areas in Xinjiang, and serves as a gene bank for wild flora and fauna. From 2023 to 2024, Ili Valley Station undertook annual ground-based tasks for ecological quality plots in grasslands, wetlands, forests, and deserts in the Ili Valley region. It has established 62 plant community monitoring plots, 207 monitoring quadrats, one bird monitoring transect, and one butterfly monitoring transect.

Subalpine mountain meadow plot

The station aims to establish itself as a pacesetter in comprehensive ecological monitoring in mountain valleys. It is equipped with an air quality automatic monitoring system and various monitoring devices such as theodolites, total stations, laser rangefinders, infrared cameras, and drones.

Mountain meadow plot; Vertical landscape of the Yili Valley

Iris ru thenica; *Malus sieversii*; *Prunus cerasifera*

Juniperus saltuaria; *Papaver pavoninum*; *Phlomoides oreophila*; *Rheum wittrockii*

Biodiversity Information

Tulipa Iliensis

Tulipa iliensis is a herbaceous perennial with bulbous structure, belonging to the genus *Tulipa* in the family *Liliaceae*. It is endemic to Xinjiang, primarily distributed in the northern slopes of the Tianshan Mountains in China, as well as in Kazakhstan and Uzbekistan. It grows at elevations between 400 and 1,100 m on piedmont plains or low mountain slopes.

Tulipa iliensis is a rare wild species under Level-II state priority conservation in China. Xinjiang is recognized as one of the world's origins of wild tulips. Currently, 14 species of wild tulips have been identified within the Tianshan region and Pamir Plateau in Xinjiang.

Malus Sieversii

Malus sieversii is a tree species in the genus *Malus* in the family *Rosaceae*, growing up to 14 m tall. It is primarily distributed in the Ili Valley and Tacheng Prefecture in Xinjiang, as well as in Tajikistan, Kazakhstan, and other countries. It typically grows at elevations between 1,100 and 1,400 m on the middle and lower slopes of mountains, or in mountain valley bottoms and river valleys.

Malus sieversii, which is under Level-II state priority conservation in China, is a remnant plant from the old-growth Mediterranean temperate deciduous forest. The famous Tianshan primitive deciduous broad-leaved wild fruit forest is dominated by *Malus sieversii*, an endangered, precious germplasm resource.

Saussurea Involucrata

Saussurea involucrata is a perennial herb in the genus *Saussurea* in the family *Asteraceae*. It mostly grows in the snowy plateau, standing tall and elegant. It resembles a snow-white lotus flower, hence the common name "Snow Lotus." Native to Xinjiang, this species is also distributed in Russia, Kazakhstan, etc. It grows at elevations between 3,600 and 4,800 m, often found in weathered zones and on rocky crevices, gravel, and sandy wetlands near the snow line. This resilient plant can germinate and grow normally within a temperature range of 5–39 °C and bloom even in harsh climatic conditions like cold winds, heavy snowfall, and intense sunlight. *Saussurea involucrata* is under Level-II state priority conservation. The entire plant can be used in traditional Chinese medicine, making it a valuable herb.

Rhodiola Quadrifida

Rhodiola quadrifida is an herbaceous perennial in the genus *Rhodiola* in the family *Crassulaceae*. It is mainly distributed in Xinjiang, China, and also found in Mongolia, Kazakhstan, and Russia. It grows at elevations between 2,300 and 5,000 m in alpine meadows, rock crevices, and scree slopes. It is under Level-II state priority conservation in China, and considered a rare and valuable medicinal plant. Both the roots and flowers can be used to address excess heat, bring down fevers, and fortify the lungs.

Part III
Charm of Wetlands

Chapter 32
Miyun Reservoir Comprehensive Ecological Quality Monitoring Station, Beijing (Wetland)

Department of Ecology and Environment Monitoring, Ministry of Ecology and Environment of the People's Republic of China and China National Environmental Monitoring Centre, *The Eye on Ecology: China's Ecological Monitoring Records (Volume 1)*, https://doi.org/10.1007/978-981-95-0047-5_32

Location of the Station

Henghe Xishan, Miyun Reservoir, Xiwengzhuang Town, Miyun District, Beijing

Lead Organization

Ecological and Environmental Monitoring Center of Beijing Municipality

Participating Organization

Beijing Miyun District Bureau of Ecology and Environment

Key Monitoring Area

Miyun Reservoir Water Source Conservation Functional Area

Latitude and Longitude

116° 54′ E

40° 29′ N

Representativeness and Typicality

- Typical warm temperate forest ecosystem
- Typical natural secondary forest in North China
- The largest reservoir in North China and the largest man-made lake in Asia.

Miyun Reservoir Comprehensive Ecological Quality Monitoring Station is located in Xiwengzhuang Town, Miyun District, Beijing, beside Miyun Reservoir, the largest source of drinking water in Beijing. The vegetation type is typical warm temperate forest ecosystem. The station is equipped with comprehensive monitoring facilities, and has established an air-space-land integrated monitoring system that combines satellite, drone, ground-based fixed and mobile monitoring. A total of 81 long-term fixed monitoring plots have been set up, focusing on monitoring the composition and structure of forest communities and biodiversity in plants and animals. Currently, a total of 2,285 species have been recorded. Among them, there are eight species under Level-I state priority conservation, 50 species under Level-II state priority conservation, 104 species under municipal priority conservation, and 130 endemic species to China. Of the 228 species of birds, *Emberiza aureola*, *Falco cherrug*, and *Grus vipio* are under Level-I state priority conservation. Over the past three years, multiple species under state priority protection have been discovered, including *Actinidia arguta*, *Cypripedium macranthos*, *Glycine soja*, *Fritillaria maximowiczii*, and *Eleutherococcus senticosus*.

Miyun Reservoir in autumn; Miyun Reservoir in winter

Miyun Reservoir; *Clanga clanga*; *Falco cherrug*; *Emberiza aureola*

Miyun Reservoir; *Otis tarda*

Batrachium pekinense; *Fritillaria maximowiczii*; *Actinidia arguta*

Biodiversity Information

Aeretes Melanopterus

The groove-toothed flying squirrel or *Aeretes melanopterus* is a species of rodent endemic to China and classified as a Vulnerable species on the Red List of China's Biodiversity. It is a close relative of squirrels, belonging to the tribe of flying squirrels (scientifically known as *Pteromyini*). However, they do not actually fly, but glide from high to low using the gliding membrane or patagium between their front and hind limbs. Their bushy tails act as a rudder during gliding. It is called the "groove-toothed flying squirrel" because there is a longitudinal "groove" on its incisors. Flying squirrels have high requirements for forest quality and are an important indicator species for forest ecosystems. Their feces are used in traditional Chinese medicine as "Wu Ling Zhi". Due to over-harvesting for Wu Ling Zhi and habitat destruction caused by human activity, these animals are nowhere to be seen in some regions. Currently, *Aeretes melanopterus* has been only recorded in Miyun District.

Discovery of New Species

Russula Paragraveolens

Russula paragraveolens is a new species discovered in 2023, belonging to the genus *Russula* in the family *Russulaceae*. Although the species in the genus *Russula* are red in color, most of them are edible, nutritious, and delicious. Among them, *Russula vinosa* is hailed as the "king of mushrooms" and was called the "mushroom for longevity" by Li Shizhen (1518–1593), a Chinese herbalist and pharmacologist of the Ming Dynasty. The species in the genus *Russula* are an essential component of China's natural drug resources and traditional Chinese herbal medicines. Many species have excellent therapeutic effects for treating lower back and leg pain, numbness in the hands and feet, discomfort in the muscles and bones, and limb spasms. Some contain anticancer substances, making them important materials for developing anticancer drugs.

Chapter 33
Bohai Rim Binhai Comprehensive Ecological Quality Monitoring Station, Tianjin (Wetland)

Department of Ecology and Environment Monitoring, Ministry of Ecology and Environment of the People's Republic of China and China National Environmental Monitoring Centre, *The Eye on Ecology: China's Ecological Monitoring Records (Volume 1)*, https://doi.org/10.1007/978-981-95-0047-5_33

Location of the Station

Qilihai Wetland, Tianjin Municipality.

Lead Organization

Tianjin University.

Participating Organizations

Ecological and Environmental Monitoring Center of Tianjin Municipality.

Tianjin Academy of Environmental Sciences.

Key Monitoring Area

Beijing-Tianjin-Hebei Estuary and Waterfront Area.

Latitude and Longitude

117° 40′ E.

38° 54′ N.

Representativeness and Typicality

- Typical coastal wetland ecosystem in northern China
- Rich biodiversity and an important species bank in China's coastal areas
- Significant impact of land-sea interactions.

Bohai Rim Binhai Comprehensive Ecological Quality Monitoring Station is located in Beidagang Wetland, Binhai New Area, and Qilihai Wetland, Ninghe District, Tianjin. This region is characterized by a typical coastal wetland ecosystem in northern China and a semi-humid continental monsoon climate, and hosts a number of natural and constructed wetlands and abundant wildlife resources.

This station focuses on studying the changes in the structure, processes, and ecological functions of coastal critical zones under the impact of climate change and human activity, as well as their relationship with resource conservation, environmental protection, and sustainable development in coastal areas. It features comprehensive observation and experimental facilities and analytical and testing platforms. The station consists of meteorological and atmospheric environment monitoring sites, material and energy flux observation sites, hydrological and geochemical observation sites, and biodiversity observation sites. It is equipped with a greenhouse gas flux observation system, standard weather observation site, automated multi-parameter water quality monitoring system, fully automated chemical analyzers, and laser spectroscopy for elemental analysis.

Beidagang Wetland

Qilihai Wetland

Reeds on Beidagang Wetland

Falco tinnunculus; *Elanus caeruleus*; *Buteo japonicus*; *Larus relictus*; *Ciconia nigra*

Numenius arquata; *Typha orientalis*; *Suaeda salsa*

Biodiversity Information

Glycine Soja

Glycine soja is an annual climbing herbaceous plant of the genus *Glycine* in the family *Fabaceae*, and a wild plant species under Level-II state priority conservation. *Glycine soja* is widely distributed across various climatic zones in China, ranging from cold temperate regions to subtropical climate regions. This plant thrives in moist environments, frequently observed in mountainous areas, along riverbanks, in wet meadows, near lakes, and in swamps or shrublands.

Its stems and small branches are slender. The leaves are trifoliate, and the racemes are typically short with small flowers. The pedicles are densely covered with long, stiff yellow hairs. The corolla of the flower is light reddish-purple or white, with nearly round banner petals. The pods are oblong, and the seeds inside are elliptical and slightly flattened. This species blooms from July to August and produces fruit from August to October.

Chapter 34
Sanjiang Plain Comprehensive Ecological Quality Monitoring Station, Heilongjiang (Wetland)

Department of Ecology and Environment Monitoring, Ministry of Ecology and Environment of the People's Republic of China and China National Environmental Monitoring Centre, *The Eye on Ecology: China's Ecological Monitoring Records (Volume 1)*, https://doi.org/10.1007/978-981-95-0047-5_34

Location of the Station

Jiansanjiang Honghe Farm, Jiamusi City, Heilongjiang Province.

Lead Organization

Northeast Institute of Geography and Agroecology, Chinese Academy of Sciences.

Key Monitoring Area

Sanjiang Plain Water Source Conservation Functional Area.

Latitude and Longitude

133° 31′ E.

47° 35′ N.

Representativeness and Typicality

- The largest marsh wetland area in China
- Key functional area of biodiversity conservation
- Typical wetland ecosystem in mid-temperate continental climate zone.

Sanjiang Plain Comprehensive Ecological Quality Monitoring Station is located in Jiansanjiang Honghe Farm, Jiamusi City, Heilongjiang Province. The Sanjiang Plain is one of the largest areas of marsh wetlands in China, featuring a wide variety of wetland types such as *Carex Linnaeus*, *Phragmites australis*, and *Deyeuxia angustifolia* marshes. It is characterized by diverse types, complex structures, and rich biodiversity. The Sanjiang Plain represents a typical wetland ecosystem in the temperate continental climate zone and exemplifies the cold-wet (long-term seasonal freezing and thawing) lowland marsh wetland types found at mid-latitudes in China. The station is dedicated to conducting comprehensive observations and studies on the structure and functions of marsh wetland ecosystems and their responses to climate change, as well as conservation of wetland biodiversity and sustainable ecosystem management. It features a large-scale field observation site and has over 150 sets of indoor analytical and field observation equipment.

Paddy field auxiliary observation site

Field observation equipment—wetland evapotranspiration observation system; Sanjiang Plain Station marsh wetland comprehensive observation site

Grus vipio; *Falco subbuteo*

Otus sunia; *Accipiter nisus*

Paris verticillata; *Caltha palustris*

Biodiversity Information

Carex Pseudocuraica

This species has relatively thick rhizomes and horizontal creeping stems that measure 1–2 m in length and are brown or dark brown. The culms emerge singly from the nodes and are arranged in rows or 2–3 clumps, reaching heights of 15–40 cm. They are flattened triangular in shape, slightly rough on the upper part, and bear leaves on the lower part. This plant grows in marshes, along lake shores, in shallow waters of wetlands, and in floodwaters of riverbanks. It is found in Heilongjiang, Jilin, Inner Mongolia, and other parts of the country.

Glyceria Spiculosa

This perennial species has long and thick rhizomes with dense fibrous roots at the nodes. The culms grow singly or in loose clumps from basal branching, and are usually erect, reaching heights of 50–120 cm with a diameter of 2.5–7 mm. Each culm typically has 9–12 nodes. This plant grows in meadows, wetlands, lakes, and marshes. It is found in Heilongjiang, Liaoning, Inner Mongolia, and other parts of the country.

Chapter 35
Taihu Comprehensive Ecological Quality Monitoring Station, Jiangsu Province (Wetland)

Department of Ecology and Environment Monitoring, Ministry of Ecology and Environment of the People's Republic of China and China National Environmental Monitoring Centre, *The Eye on Ecology: China's Ecological Monitoring Records (Volume 1)*, https://doi.org/10.1007/978-981-95-0047-5_35

Location of the Station

North of Tanhua Temple, Huantaihu Avenue, Wuzhong District, Suzhou City, Jiangsu Province.

Lead Organization

Ecological and Environmental Monitoring Center of Jiangsu Province.

Participating Organization

Nanjing Institute of Geography and Limnology, Chinese Academy of Sciences.

Satellite Application Center for Ecology and Environment, Ministry of Ecology and Environment.

Key Monitoring Area

Other key ecosystem service zone.

Latitude and Longitude

120°22′ E.

31°13′ N.

Representativeness and Typicality

- Five types of ecosystems: urban–rural ecosystem, forest ecosystem, wetland ecosystem, farmland ecosystem, and water body
- Regionally representative biodiversity and vegetation ecosystems
- China's largest recursive island that lies within a lake.

Taihu Ecosystem Research and Field Observation Station is located on the shores of the Taihu Lake and consists of the Taihu Lake Water Quality and Cyanobacteria Comprehensive Observation Station (Suzhou) and the Taihu Ecosystem Research and Field Observation Station (Wuxi). The station is dedicated to conducting comprehensive field observations of water quality and cyanobacteria in the Taihu Lake, monitoring the total amount of major water pollutants and risk warnings, and studying the risks associated with persistent organic pollutants. It is capable of ecological and environmental monitoring over multiple elements and types, such as water quality, cyanobacteria bloom, aquatic organisms, terrestrial biodiversity, ecological remote sensing, soil, and air. Each year, it generates over 110,000 entries of data from field observations and laboratory analysis.

Taihu Lake Water Quality and Cyanobacteria Comprehensive Observation Station, Jiangsu Province

Taihu Ecosystem Research and Field Observation Station (Wuxi), Chinese Academy of Sciences

Hydrophasianus chirurgus; *Nettapus coromandelianus*

Leiothrix lutea

Cygnus cygnus; *Egretta eulophotes*; *Aythya baeri*; *Cygnus columbianus*

Biodiversity Information

Psilotum Nudum

Psilotum nudum, the whisk fern, belongs to the genus *Psilotum* in the family *Psilotaceae*. It is an epiphytic perennial herb that grows on tree trunks or rocks. It has creeping underground stems with dichotomous branching and only has hair-like water-absorbing structures or rhizoids. The aerial stems are erect or pendulous, reaching heights of 15–80 cm. This plant thrives in warm, humid, and semi-shaded environments. *Psilotum nudum* has slender branches. The entire plant can be used for medicinal purposes. It is a relict species that has existed since the Devonian period over 300 million years ago, making it one of the oldest and most primitive vascular plants in existence today. It is also one of the oldest and most primitive terrestrial higher plants.

Ceratopteris Thalictroides

When young, the plant looks green and succulent, with tender and soft stems. Due to varying water conditions, its morphology can differ significantly, with some reaching heights of up to 70 cm. The rhizome of this fern is short and erect, supported by a cluster of thick roots that anchor firmly in the mud. The plant thrives in ponds, paddy fields, and ditches, and may even float on the surface of deeper waters. This species can be used for medicinal purposes. The stems and leaves are traditionally used to treat fetal toxicity and get rid of phlegm and mucus. The tender leaves can also be consumed as vegetables.

Chapter 36
Changzhou Plain Water Network Station, Jiangsu Province (Wetland)

Department of Ecology and Environment Monitoring, Ministry of Ecology and Environment of the People's Republic of China and China National Environmental Monitoring Centre, *The Eye on Ecology: China's Ecological Monitoring Records (Volume 1)*, https://doi.org/10.1007/978-981-95-0047-5_36

Location of the Station

Gehu Lake and Tianmu Lake, Changzhou City, Jiangsu Province.

Lead Organization

Ecological and Environmental Monitoring Center of Changzhou.

Participating Organization

Nanjing Institute of Geography and Limnology, Chinese Academy of Sciences.

Key Monitoring Area

Other key ecosystem service zone.

Latitude and Longitude

119° 52′ E.

31° 38′ N.

119° 24′ E.

31° 15′ N.

Representativeness and Typicality

- Densely distributed water networks and river networks
- High biodiversity
- Key water source conservation functional area.

Changzhou Plain Water Network Station is located in the Taihu Lake Water Source Conservation Functional Area, Jiangsu Province. The Taohu-Gehu Wetland and Yixing-Liyang Low Mountains and Hills where the station is located are the biodiversity center of regional aquatic ecosystems and terrestrial ecosystems, respectively. The region features diverse landform types, such as low mountains, hills, plains, polders, lakes, ponds, rivers, and reservoirs. The station has comprehensive monitoring facilities, with 20 test ponds for ecological monitoring, and over 40 sets of test pond monitoring devices, AI-powered bird monitoring devices, carbon flux observation devices, and video surveillance systems. Additionally, it has 12 sets of water vapor/carbon flux monitoring instruments, and more than 10 automated hydrological and water quality stations with algal classification capabilities for rivers and lakes. These facilities enable comprehensive online monitoring of water quality, algae, environmental DNA, meteorology, carbon flux, soil, and bird observations. The station is also equipped with remote data transmission devices such as drones, unmanned boats, and vehicles for mobile monitoring, capable of conducting multi-element monitoring.

Tianmu Lake Wetland

Maoshan Mountain

Gehu Lake

A paradise for birds

Elanus caeruleus; *Hydrophasianus chirurgus*

Biodiversity Information

Shaniodendron Subaequal

Shaniodendron subaequal, also known as *Parrotia subaequalis*, is a wild plant under Level-I state priority conservation. It can grow up to 4–5 m in height. The leaves are green on the upper surface and light brown on the underside. It has a head-like inflorescence, and the capsules are shaped nearly round. The seeds are spindle-shaped, pointed at both ends, shiny, and brown. *Shaniodendron subaequal* is one of the oldest angiosperms, discovered in fossils dating back to the Cretaceous period of the Mesozoic era, approximately 67 million years ago. Like the gymnosperms *Ginkgo biloba* and *Metasequoia glyptostroboide*s, it is a "living fossil" species that has been rediscovered in China. This plant has extremely high habitat requirements, and is only found in limited numbers in the county-level cities Liyang and Yixing of Jiangsu Province, Jinzhai County of Anhui Province, and Anji County of Zhejiang Province.

Novaculina Chinensis

Novaculina chinensis belongs to the family *Solecurtidae* in the class *Bivalvia* of the order *Venerida.* It is a rare freshwater bivalve under Level-II state priority conservation and has high habitat requirements, typically inhabiting muddy or sandy bottoms of rivers and lakes. This species requires sandy soil for reproduction and thrives in clear, transparent water. Due to these specific ecological conditions, the population of *Novaculina chinensis* is relatively small in China, so it is classified as a rare species. It is found in the main lakes of Taohu-Gehu Wetland in Changzhou, including the Changdang Lake and the Gehu Lake.

Chapter 37
Yancheng Binhai Comprehensive Ecological Quality Monitoring Station, Jiangsu Province (Wetland)

Department of Ecology and Environment Monitoring, Ministry of Ecology and Environment of the People's Republic of China and China National Environmental Monitoring Centre, *The Eye on Ecology: China's Ecological Monitoring Records (Volume 1)*, https://doi.org/10.1007/978-981-95-0047-5_37

Location of the Station

Dongchuan Wetland, Dafeng District, Yancheng City, Jiangsu Province.

Lead Organization

Jiangsu Dafeng Elk National Nature Reserve Administration.

Participating Organization

Ecological and Environmental Monitoring Center of Yancheng.

Key Monitoring Area

North Jiangsu Coastal Wetland Biodiversity Reserve.

Latitude and Longitude

120°48′E.

32°59′N.

Representativeness and Typicality

- Various vegetation types, including halophytes, reeds, wetland meadows and mangrove plants
- Typical wetland ecosystem featuring diverse habitats such as woodlands, grasslands, saline bare land, mudflats, and marshes.

Yancheng Binhai Comprehensive Ecological Quality Monitoring Station is located in Dafeng District, Yancheng City, Jiangsu Province. It lies on the Coast of Yellow Sea-Bohai Gulf of China (Phase I), a World Natural Heritage site. This region is characterized as a typical alluvial plain with tidal flats, features a transitional climate zone between marine and monsoon climates, and is located in the transition zone between subtropical and warm temperate regions.

The station is dedicated to biodiversity monitoring, with particular emphasis on mammal diversity, bird diversity, wetland plots, environmental DNA, and ecological remote sensing. It also pays attention to species composition, structure, functions, and stress. The station is well-equipped with two test sites, 83 monitoring points, and four observation towers. It conducts real-time monitoring of the activity patterns, populations, and distribution range of *Elaphurus davidianus* in the wild, as well as regular bird monitoring and banding activities. The region has high biodiversity, hosting 322 species of plants, 24 species of mammals, 283 species of birds, 27 species of amphibians and reptiles, and 233 species of insects.

Landscape around the station

Platalea minor; *Asio tus*; *Phasianus colchicus*; *Ciconia boyciana*

Pennisetum alopecuroides; *Artemisia halodendron*

Biodiversity Information

Elaphurus Davidianus

Elaphurus davidianus, also known as the milu deer, belongs to the genus *Elaphurus* in the family *Cervidae* and is nicknamed "sibuxiang" (literally meaning "like none of the four") for its unique appearance: a long face similar to that of a horse, antlers with multiple tines like those of a deer, wide hooves resembling those of a cow, and a long tail similar to that of a donkey. It is a rare species endemic to China and under Level-I state priority conservation. It can reach up to 2 m in height and approximately 300 kg in weight. Males grow antlers, while females do not. The summer coat of hair is relatively sparse and reddish-brown in color, and the winter coat is thicker and turns pale yellowish-green like the color of withered grass. Native to the marshes in the middle and lower reaches of the Yangtze River, the milu deer became extinct in China in the early twentieth century. In 1986, 39 milu deer were reintroduced to Dafeng County to restore the population, and the number of deer has now grown to 8,216.

Grus Japonensis

Grus japonensis belongs to the genus *Grus* in the family *Gruidae*, is a slender and elegant bird with a bold black and white plumage. It is named for the distinctive red crown on its head. Primarily distributed in Northeast China, the Russian Far East, and Japan, it is an important migratory bird inhabiting wetlands, marshes, and rice paddies, where it feeds mainly on aquatic plants, insects, and small fish. This species is known for graceful dance and melodious call, often performing dance routines during courtship and social interactions.

Discovery of New Species

Hypsipetes Amaurotis

Hypsipetes amaurotis belongs to the family *Pycnonotidae* of the order *Passeriformes*. The top of its head to the back of its neck is gray, with chestnut-colored ear coverts and sides of the neck. The crown and nape are gray, and the wings and tail are brownish-gray. The throat and chest have light longitudinal stripes, and the belly is basically white, with the flanks marked by gray spots. The rump is marked with black and white bars. This bird mainly feeds on fruits and seeds from trees and shrubs. It inhabits low mountain broad-leaved forests, mixed forests, and forest edges, typically distributed in Japan, North Korea, the Philippines, Northeast China, and the Bohai Rim region. It is generally considered a rare bird in Jiangsu Province. It was discovered in the seawall forest of the reserve on March 2, 2024, and recorded for the first time in Yancheng.

Chapter 38
Poyang Lake Station, Jiangxi Province (Wetland)

Department of Ecology and Environment Monitoring, Ministry of Ecology and Environment of the People's Republic of China and China National Environmental Monitoring Centre, *The Eye on Ecology: China's Ecological Monitoring Records (Volume 1)*, https://doi.org/10.1007/978-981-95-0047-5_38

Location of the Station

Lushan County-level City, Jiujiang City and Xinjian District, Nanchang City, Jiangxi Province.

Lead Organization

Ecological and Environmental Monitoring Center of Jiangxi Province.

Participating Organizations

Poyang Lake Wetland Research Station, Nanjing Institute of Geography and Limnology, Chinese Academy of Sciences.

Satellite Application Center for Ecology and Environment, Ministry of Ecology and Environment.

Jiangxi Academy of Eco-Environmental Sciences and Planning.

Jiangxi Normal University.

Key Monitoring Area

Poyang Lake Flood Regulation and Storage and Biodiversity Conservation.

Latitude and Longitude

116°03′E.

29°26′N.

116°20′E

28°53′N.

Representativeness and Typicality

- Large shallow lake ecosystems in typical humid subtropical monsoon climate zones
- Naturally connected floodplain lake wetland ecosystem in the middle reaches of the Yangtze River
- Unique landscape and environmental heterogeneity as a result of seasonal water level changes
- Regionally representative plant and animal diversity.

Poyang Lake Station in Jiangxi Province includes two locations: Lushan Station and Nanji Station. The station focuses on studying the eco-environment factors of Poyang Lake wetlands, hydrogeochemical elements and water quality of lake wetlands, and aquatic environments. It conducts long-term monitoring and comprehensive analysis of the current status and evolution patterns of the Poyang Lake ecosystem.

The station is equipped with a standard meteorological observation field, an experimental field for wetland ecological control, and multi-element monitoring transects in tidal flat wetlands. It also has a number of monitoring devices, including multi-parameter water quality analyzers, acoustic Doppler current profilers (ADCPs), high-performance ground penetrating radar (GPR), a water vapor flux monitoring system, soil laser particle size analyzers, and large drones. These facilities enable the station to conduct air-space-land integrated monitoring and simultaneous online monitoring of water quantity, quality, and ecology.

A panoramic view of Poyang Lake Station

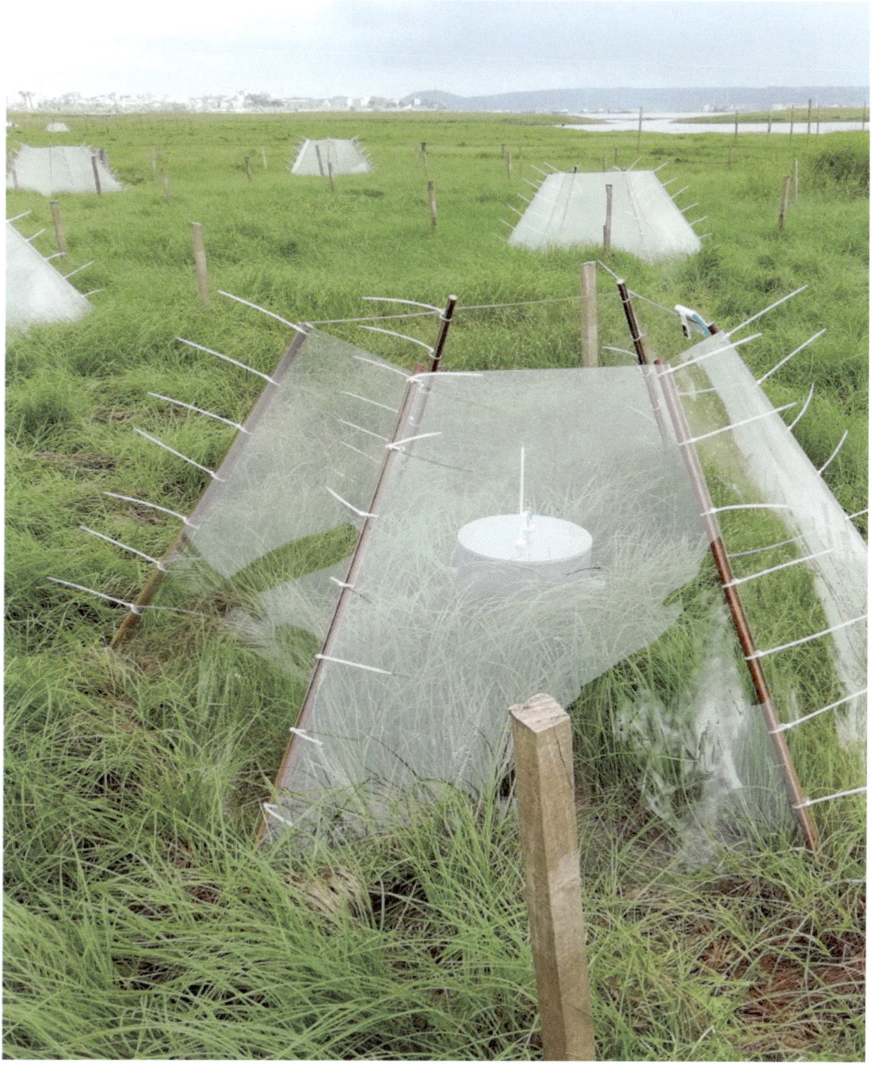

Tidal flat experimental plot (flux tower)

Ciconia boyciana; *Grus monachal*

Anas zonorhyncha; *Platalea leucorodia*; *Anscr cygnoidcs*; *Grus vipio*

Biodiversity Information

Artemisia Selengensis

Artemisia selengensis is one of the representative wetland plant species in the Poyang Lake region, and its distribution can be seen across various sandbanks in the lake area. It is a perennial herb with a distinctive aromatic scent. The stems are few or solitary, standing 60–150 cm tall, initially greenish-brown and turning purplish-red as they mature, without hair and featuring prominent longitudinal ridges. Commonly found in strip or patch formations between the *Carex cinerascens* community and the reed (*Miscanthus sacchariflorus*) community, it forms dominant plant populations or can also appear as an accompanying species within the grass and reed communities The entire plant is rich in essential oils, providing unique fragrance. It is used in traditional Chinese medicine for hemostasis, anti-inflammation, cough suppression, and phlegm resolution, and the tender stems are edible, making it an economically important wild plant in the region.

Grus Leucogeranus

Grus leucogeranus, also known as the Siberian Crane or Black-sleeved Crane, is a large wading bird designated as the "Provincial Bird" of Jiangxi and under Level-I state priority conservation in China. Globally, it is considered a critically endangered species and recognized as the "living fossil" of birds. According to statistics from the International Union for Conservation of Nature, the global population of *Grus leucogeranus* is less than 4,000, with 98% of them wintering in Poyang Lake. It is the most habitat-specialized crane species, whose survival is closely tied to specific environmental conditions, particularly shallow wetlands. It primarily feeds on the roots and stems of aquatic plants and also consumes small quantities of clams, fish, and snails.

Chapter 39
Honghu Comprehensive Ecological Quality Monitoring Station, Hubei (Wetland)

Department of Ecology and Environment Monitoring, Ministry of Ecology and Environment of the People's Republic of China and China National Environmental Monitoring Centre, *The Eye on Ecology: China's Ecological Monitoring Records (Volume 1)*, https://doi.org/10.1007/978-981-95-0047-5_39

Location of the Station

Honghu County-level City, Jingzhou City, Hubei Province.

Lead Organization

Innovation Academy for Precision Measurement Science and Technology, Chinese Academy of Sciences.

Participating Organizations

Ecological and Environmental Monitoring Center of Hubei Province.

Center of Ecological and Environment Monitoring and Research, Yangtze River Basin Eco-environmental Supervision and Administration Agency, Ministry of Ecology and Environment.

Key Monitoring Area

Honghu Lake Basin Water Source Conservation and Biodiversity Conservation Functional Area.

Latitude and Longitude

113°29′E.

29°55′N.

Representativeness and Typicality

- Unique lake-farmland integrated wetland ecosystem
- The largest lake in Hubei, known as the "province of thousands of lakes"
- Wetland of international importance and national nature reserve
- Priority area for biodiversity conservation designated by the World Wildlife Fund (WWF).

Honghu Comprehensive Ecological Quality Monitoring Station is located in the lakeside area of Honghu County-level City, Jingzhou, Hubei Province, which is part of the vital wetland ecological zone in the middle reaches of the Yangtze River. This region serves as an important habitat and wintering area for migratory waterfowl, and is hailed as the "gene bank" of wetland species in the middle reaches of the Yangtze River.

The station focuses on studying the structural and functional evolution of wetland ecosystems on lake plains under the impact of climate change and human activity. It has established one comprehensive meteorological observation site, 30 wetland water quality observation points, 33 submerged vegetation observation points, six lakeshore vegetation observation points, and 12 benthic and planktonic organisms observation points. The station is equipped with over 120 sets of monitoring equipment, including meteorological observation instruments, spectrometers, canopy analyzers, multibeam echosounders, and acoustic Doppler current profilers. This region has great biodiversity. Surveys have identified 286 species of vascular plants belonging to 197 genera of 70 families, including *Glycine soja*, *Nelumbo nucifera*, *Trapa incisa*, and *Ceratopteris thalictroides*. There are 288 species of phytoplankton in seven phyla, 205 species of terrestrial vertebrates belonging to 72 families of 28 orders across four classes, 123 species of mayflies, and 36 species of benthic animals. It also hosts 87 species of fish, such as *Hoplobatrachus chinensis*, *Mauremys reevesii*, *Grus monacha*, and *Elaphodus cephalophus*.

Constructed wetland ecosystem in agricultural fields

Nelumbo nucifera; *Euryale ferox–Zizania latifolia* community; *Nelumbo nucifera–Salvinia natans* community

Chlidonias hybrida

Bubulcus ibis

Hydrophasianus chirurgus

Biodiversity Information

Chinese Salamander

The Chinese salamander (*Hynobius chinensis*) is a unique amphibian endemic to China and under Level-I state priority conservation. Due to its critically endangered status, the species was included in *China Red Data Book of Endangered Animals* in 1986, alongside the giant panda (*Ailuropoda melanoleuca*). It was thought to be extinct in the 1990s. Fortunately, Chinese scholars rediscovered the species at its type locality in 2005, 116 years after its scientific naming.

Chapter 40
Fuxian Lake Plateau Lake Station, Yunnan Province (Wetland)

Department of Ecology and Environment Monitoring, Ministry of Ecology and Environment of the People's Republic of China and China National Environmental Monitoring Centre, *The Eye on Ecology: China's Ecological Monitoring Records (Volume 1)*, https://doi.org/10.1007/978-981-95-0047-5_40

Location of the Station

North shore of Fuxian Lake, Yousuo Town, Chengjiang County-level City, Yuxi City, Yunnan Province.

Lead Organization

Ecological and Environmental Monitoring Center of Yunnan Province.

Participating Organizations

Nanjing Institute of Geography and Limnology, Chinese Academy of Sciences.

Fuxian Lake Plateau Deepwater Lake Research Station.

Ecological and Environmental Monitoring Station in Yuxi City, Department of Ecology and Environment of Yunnan Province.

Key Monitoring Area

Other key ecosystem service zone.

Latitude and Longitude

102°54′E.

24°38′N.

Representativeness and Typicality

- The second deepest freshwater lake in China
- China's strategic water resource
- Unique biota, with plenty of endemic fish species and prominent sympatric speciation
- An ideal region for studying the aquatic environments of plateau deepwater lakes.

Fuxian Lake Plateau Lake Station is situated on the picturesque northern shore of the Fuxian Lake in Yunnan, at an elevation of 1,728 m. The lake belongs to the Nanpan River Basin within the Xijiang Water System and is the largest oligotrophic deep lake in China with the best water quality and the highest storage capacity among inland freshwater lakes. Its overall water quality has long stood at Grade I. The maximum depth of the Fuxian Lake is 158.9 m, and its highest storage capacity is 20.62 billion cubic meters, accounting for 9.2% of the total freshwater storage in lakes across China. The station focuses on studying the three lakes of the central Yunnan Plateau (Fuxian Lake, Xingyun Lake, and Qilu Lake). It is equipped with air-space-land integrated monitoring facilities such as research vessels, an AI-based water quality laboratory, an environmental DNA laboratory, aquatic organism microscopy rooms, underwater robots, automated drones, an automated water quality monitoring network, and an intelligently triggered sampling network for non-point source pollution in the basin.

Stunning scenery of Fuxian Lake

Gekko reevesii; *Lophura nycthemera*

Merops viridis; *Strix nivicolum*; *Merops philippinus*; *Psansomus dalhousiae*

Paphiopedilum gratrixianum; *Paphiopedilum wenshanense*; *Renanthera imschootiana*

Biodiversity Information

Sinocyclocheilus Tingi

Sinocyclocheilus tingi belongs to the genus *Sinocyclocheilus* in the family *Cyprinidae* and is classified as a Vulnerable species under Level-II state priority conservation. It is a small- to medium-sized fish, characterized by its elongated and laterally compressed body shape and a pointed snout. It is mainly distributed in the shallow coastal areas and mid-lower layers of open waters in Fuxian Lake, preferring rocky areas with spring water or near caves. It prefers to live in groups and inhabits burrows, is a slow swimmer, and primarily feeds on small fish and shrimp. It spawns in winter, and the peak breeding season typically spans the low-temperature period from January to March around the Chinese New Year.

Anabarilius Grahami

Anabarilius grahami, commonly known as the Kanglang fish, belongs to the genus *Anabarilius* in the family *Cyprinidae*. Endemic to Fuxian Lake in Yunnan, it is classified as a Vulnerable species. It has a small and slender body that is silver-white on the sides and belly, a pointed snout, and no barbels. The upper and lower edges of the eyes are red, and the fins are grayish-white. It is a top to mid-dwelling fish species that prefers oxygen-rich open waters, and is known for its rapid swimming and strong resistance to waves. This fish species begins to spawn in mid-to-late April.

Part IV
Melody in the Wilderness

Chapter 41
Kubuqi Comprehensive Ecological Quality Monitoring Station, Inner Mongolia (Desert)

Department of Ecology and Environment Monitoring, Ministry of Ecology and Environment of the People's Republic of China and China National Environmental Monitoring Centre, *The Eye on Ecology: China's Ecological Monitoring Records (Volume 1)*, https://doi.org/10.1007/978-981-95-0047-5_41

Location of the Station

Guannianfang Village, Shulinzhao Town, Dalad Banner, Ordos City, Inner Mongolia Autonomous Region

Lead Organization

Environmental Monitoring Center of Inner Mongolia

Participating Organization

Institute of Grassland Research, Chinese Academy of Agricultural Sciences

Key Monitoring Area

Ordos Plateau Functional Area of Wind Prevention and Sand Fixation

Latitude and Longitude

110°00′ E

40° 19′ N

Representativeness and Typicality

- Transition zone between semi-arid grassland and arid semi-desert (desert grassland)
- Typical sub-shrub *Artemisia* vegetation in sandy land with *Artemisia ordosica* as the constructive species.

Located in the Kubuqi Desert, China's seventh largest desert, Kubuqi Comprehensive Ecological Quality Monitoring Station is part of the Ordos Plateau Functional Area of Wind Prevention and Sand Fixation. The site encompasses various landforms such as fixed sand dunes, semi-fixed sand dunes, mobile sand dunes, and inter-dune lowlands. With an intricate and varied ecosystem, this region is an eco-sensitive area within the Yellow River Basin and a priority area for building the ecological security shields in northern China. The dominant vegetation includes desert-adapted plants. The constructive species on fixed and semi-fixed sand dunes is *Artemisia ordosica*, and species distributed on mobile sand dunes include *Artemisia blepharolepis*, *Agriophyllum pungens*, and *Inula salsoloides*.

This station focuses on the monitoring and research of sandy land ecosystems and grassland desertification, equipped with over 80 sets of field ecological positioning observation and indoor experimental instruments. It conducts observation and research on atmosphere, vegetation, soil, hydrology, etc. The region harbors 124 species of vascular plants belonging to 78 genera in 28 families, including one species of fern, one species of gymnosperm, and 122 species of angiosperms in 76 genera of 26 families.

A distant view of Kubuqi Station

Landscape of Kubuqi Desert

Atraphaxis bracteata; *Aster altaicus*; *Orobanche coerulescens*; *Allium mongolicum*; *Euphorbia esula*; *Calligonum mongolicum*; *Caragana liouana*

Artemisia blepharolepis; *Phragmites australis*; *Psammochloa villosa*; *Artemisia ordosica*; *Pugionium cornutum*; *Agropyron desertorum*

Biodiversity Information

Ammopiptanthus Mongolicus

Ammopiptanthus mongolicus is an evergreen shrub belonging to the genus *Ammopiptanthus* in the family *Fabaceae*, growing up to 2 m in height. It is found in regions such as Inner Mongolia, Ningxia, and Gansu, and known for its exceptional sand-fixing abilities. It is one of the first rare and endangered species under state priority conservation, and a Tertiary relict plant endemic to central Asia. Notably, it is the only evergreen broad-leaved shrub adapted to extreme drought conditions in the desert regions of Northwest China. This species is included in China's List of Wild Plants Under State Priority Conservation (Level II), and classified as a Vulnerable species on the IUCN Red List of Threatened Species. *Ammopiptanthus mongolicus* is known as a "living fossil" that has undergone significant changes since ancient geological times, and also provides important evidence for studying ancient climate changes and geographical environmental transformations.

Agropyron Mongolicum

Agropyron mongolicum, a perennial grass in the family *Poaceae*, grows in loose tufts and can reach up to 60 cm. Distributed across Inner Mongolia, Shanxi, Shaanxi, and Gansu, this species thrives in dry grasslands and sandy areas. As a typical xerophyte, it is highly resistant to drought, cold, and windblown sand. It has relatively low soil requirements, making it an excellent choice for sand fixation. Ideal for artificial reseeding in degraded pastures, it is also a prime forage for improving sandy grasslands. This species is included in China's List of Wild Plants Under State Priority Conservation (Level II).

Chapter 42

Mu Us Comprehensive Ecological Quality Monitoring Station, Inner Mongolia (Desert)

Department of Ecology and Environment Monitoring, Ministry of Ecology and Environment of the People's Republic of China and China National Environmental Monitoring Centre, *The Eye on Ecology: China's Ecological Monitoring Records (Volume 1)*, https://doi.org/10.1007/978-981-95-0047-5_42

Location of the Station

Naoerhao Village, Nalintaohai Town, Ejin Horo Banner, Ordos City, Inner Mongolia.

Lead Organization

Environmental Monitoring Center of Inner Mongolia.

Participating Organization

Institute of Botany, Chinese Academy of Sciences.

Key Monitoring Area

Ordos Plateau Functional Area of Wind Prevention and Sand Fixation.

Latitude and Longitude

110°11′E.

39°29′N.

Representativeness and Typicality

- Key functional area of wind prevention and sand fixation
- Typical shrub grassland vegetation in temperate grassland sandy areas.

Mu Us Comprehensive Ecological Quality Monitoring Station is situated in Ejin Horo Banner, Ordos City, Inner Mongolia. This region has a temperate continental semi-arid climate, characterized by four distinct seasons. The primary vegetation types include sand-fixing shrublands on fixed and semi-fixed sand dunes.

The station focuses on studying the methods and technologies for controlling mobile sand dunes, plant adaptability, and ecological conservation in coordination with socioeconomic development. It has pioneered a "three-circle" ecological conservation and production model for sustainable management of desertified land. The station is equipped with advanced field observation devices for soil, vegetation, meteorology, soil erosion, and remote sensing, including automated soil moisture monitoring instruments, automated plant growth rhythm monitoring systems, meteorological radiation observation equipment, multi-channel surface canopy spectral measurement systems, and runoff measurement systems.

Mu Us Station; Sand-tolerant vegetable cultivation

Artemisia ordosica community

Juniperus sabina; Salix cheilophila; Agriophyllum pungens; Caragana korshinskii

Biodiversity Information

Psammochloa Villosa

Chapter 43
Hunshandake Comprehensive Ecological Quality Monitoring Station, Inner Mongolia (Desert)

Department of Ecology and Environment Monitoring, Ministry of Ecology and Environment of the People's Republic of China and China National Environmental Monitoring Centre, *The Eye on Ecology: China's Ecological Monitoring Records (Volume 1)*, https://doi.org/10.1007/978-981-95-0047-5_43

Location of the Station

Arilin Chagantala, Zhenglan Banner, Xilin Gol League, Inner Mongolia.

Lead Organization

Inner Mongolia Comprehensive Ecological Quality Monitoring Station.

Participating Organization

Institute of Botany, Chinese Academy of Sciences.

Key Monitoring Area

Hunshandake Desertification Control Area.

Latitude and Longitude

116° 1′ E.

42° 54′ N.

Representativeness and Typicality

- Typical elm sparse forest, shrub and grassland landscapes
- Regionally representative landscape and biodiversity
- Source of sandstorms and important functional area of wind prevention and sand fixation for the Beijing-Tianjin region.

Hunshandake Comprehensive Ecological Quality Monitoring Station is located in Arilin Chagantala, Zhenglan Banner, Xilin Gol League, Inner Mongolia, in the heart of the Hunshandake Sandy Land, one of the four major sandy lands in northern China. Its southern edge is just 180 km from Beijing, making it a critical area in the Beijing-Tianjin Sandstorm Source Control Project and a functional area of vital importance for wind prevention and sand fixation. It encompasses various typical habitats, including fixed sand dunes, semi-fixed sand dunes, mobile sand dunes, meadows, and wetlands. Additionally, it features a typical landscape of sparse elm forests on the grassland, and has rich biodiversity, hosting an abundance of species from families such as *Asteraceae*, *Poaceae*, *Fabaceae*, *Rosaceae*, and *Lamiaceae*.

This station focuses on studying the mechanisms of vegetation recovery in degraded sandy grassland ecosystems and exploring effective approaches for the adaptive management of sandy grassland ecosystems, aiming to enhance ecological, economic, and social benefits in a coordinated manner.

Inside Hunshandake Station

Landscape around Hunshandake Station

Capreolus pygargus

Agriophyllum pungens; Mobile sand dunes after control efforts

Ulmus pumila under a double rainbow

Biodiversity Information

Ulmus Pumila

Ulmus pumila is a deciduous tree belonging to the genus *Ulmus* in the family *Ulmaceae* and can grow up to 20–25 m tall. It has gray-black bark with longitudinal fissures. The crown is nearly round or oval in shape, with spreading branches. The flowering period is from March to April, the fruiting period from April to May, and its leaves fall in October. It is a dominant species in the sandy lands of northern China, especially typical in the Hunshandake Sandy Land, where it is most widely distributed in the Zhenglan Banner of Xilin Gol League.

Pugionium Cornutum

Pugionium cornutum is a biennial herb of the *Pugionium* genus in the *Brassicaceae* family and is endemic to China. The genus is unique to the sandy regions of the

Mongolian Plateau and includes two species: *Pugionium cornutum* and *Pugionium dolabratum*, of which the latter was listed as a rare and endangered plant in the book *Rare and Endangered Plants in Inner Mongolia* in 1992. Plants of the genus *Pugionium* have multiple functions including culinary use, forage, medicinal use, health benefits, wind prevention, and sand fixation. They are sand-tolerant plants that are both medicinal and edible, holding high value in terms of development, utilization, and eco-environmental protection. Plants of the genus *Pugionium* are found growing on semi-fixed or mobile sand dunes and are exclusively distributed in certain sandy regions of China, such as the Horqin, Hunshandake, Kubuqi, Mu Us, and Hedong of Ningxia.

Chapter 44
Heihe River Middle Reaches Oasis Station, Gansu (Desert)

Department of Ecology and Environment Monitoring, Ministry of Ecology and Environment of the People's Republic of China and China National Environmental Monitoring Centre, *The Eye on Ecology: China's Ecological Monitoring Records (Volume 1)*, https://doi.org/10.1007/978-981-95-0047-5_44

Location of the Station

2 km south of Dangzhai Town, Ganzhou District, Zhangye City, Gansu Province.

Lead Organization

Northwest Institute of Eco-Environment and Resources, Chinese Academy of Sciences.

Participating Organizations

Northwest Institute of Eco-Environment and Resources, Chinese Academy of Sciences.

Heihe Remote Sensing Research Station.

Key Monitoring Area

Hexi Corridor Key Functional Area of Wind Prevention and Sand Fixation.

Latitude and Longitude

100° 29′ E.

38° 50′ N.

Representativeness and Typicality

- Mixed desert and oasis ecosystems in arid areas
- Arid and semi-arid eco-sensitive area in Northwest China
- Well-developed irrigated agriculture in the oasis
- Regionally representative arid vegetation.

Heihe River Middle Reaches Oasis Station ("Heihe Station" for short) is located in Ganzhou District, Zhangye City, Gansu Province, at an elevation of 1,525 m. It features a typical temperate continental climate with four distinct seasons and a dry climate, with annual precipitation of approximately 100 mm. The region is home to typical desert and oasis ecosystems in the northwestern part of the country. The station is equipped with a comprehensive monitoring infrastructure, including an eddy covariance (EC) system, large aperture scintillometer (LAS), automated weather station (AWS), wireless sensor network (WSN), and cosmic-ray soil moisture observation system. It has the capability of integrated space-air-land monitoring and features observation fields for desert ecosystems, oasis farmlands, and oasis wetlands.

Heihe Station

Huazhaizi Desert Station

Biodiversity Information

Reaumuria Songarica

Reaumuria songarica is a drought-tolerant shrub belonging to the genus *Reaumuria* in the family *Tamaricaceae*, typically found in dry and barren Gobi and desert areas. This plant exhibits remarkable adaptability, capable of surviving in arid regions at higher elevations and withstanding extreme climatic conditions. It is primarily distributed in the northwest regions of China, including Gansu, Qinghai, and Inner Mongolia. The plant is small, generally 10 to 30 cm tall, and features a well-developed root system.

Reaumuria songarica

Caroxylon Passerinum

Caroxylon passerinum is a perennial herb that typically grows in arid and saline soils, exhibiting strong adaptability. It typically grows in desert and semi-desert areas at lower elevations, and is tolerant of severe drought conditions. It is primarily distributed in the desert regions of Northwest China, including Gansu, Ningxia, and Xinjiang.

Caroxylon passerinum

Chapter 45
Jungar Comprehensive Ecological Quality Monitoring Station, Xinjiang (Desert)

Department of Ecology and Environment Monitoring, Ministry of Ecology and Environment of the People's Republic of China and China National Environmental Monitoring Centre, *The Eye on Ecology: China's Ecological Monitoring Records (Volume 1)*, https://doi.org/10.1007/978-981-95-0047-5_45

Location of the Station

Wujiaqu City, Xinjiang Production and Construction Corps Sixth Agricultural Division.

Lead Organization

Xinjiang Academy of Environmental Protection Science.

Key Monitoring Area

Gobi-Desert-Oasis Region in Jungar Basin.

Latitude and Longitude

87° 31′ E.

44° 30′ N.

Representativeness and Typicality

- Desert-oasis ecotone
- Regionally representative biodiversity and vegetation ecosystems.

Established in 1984, Jungar Station is located on the desert-oasis ecotone at the southern edge of the Jungar Basin. The region has a temperate continental arid climate characterized by short, hot summers and long, cold winters with drastic temperature fluctuations, significant diurnal temperature variations, and scarce and unevenly distributed precipitation. It has strong evaporation, intense solar radiation, and sparse vegetation, dominated by desert plants such as *Tamarix chinensis*, *Haloxylon persicum*, and *Calligonum mongolicum.*

The station focuses on the observation of the typical desert-oasis ecotone and the complete vertical vegetation spectrum of inland rivers in the arid region. It monitors changes in ecological quality, systematically identifies key environmental issues such as the structural succession of typical ecosystems and oasis development, and studies the impact of human activity on the eco-environment of the ecotone, environmental pollution mechanisms, and other topics, and accumulates ecosystem monitoring data.

The landscape

Echinops sphaerocephalus; *Karelinia caspia*; *Pyankovia roborowskii*; *Caragana halodendron*

Gazella subgutturosa; *Vulpes vulpes*; *Phrynocephalus melanurus*

Athene noctua; *Psammophis lineolatus*; *Rhombomys opimus*

Biodiversity Information

Tamarix Chinensis

Tamarix chinensis is a shrub or small tree belonging to the genus *Tamarix* in the family *Tamaricaceae*, growing up to 1.5 to 6 m hall. It is one of the dominant species in desert areas, known for its excellent fuelwood and afforestation properties, as well as its sand-fixing abilities. Its branches and leaves are used in traditional Chinese medicine to reduce fever, induce rash, expel wind and dampness, and increase urine output. The tender branches can also be used as livestock feed.

Vormela Peregusna

Vormela peregusna belongs to the genus *Vormela* in the family *Mustelidae.* It has a slender and uniform body, elliptical ears, and short, sturdy limbs. The dorsal side of its body is yellowish-white with numerous brown or pinkish-brown spots, while its abdomen is dark brown. This species inhabits deserts and semi-desert grasslands, where it digs burrows for shelter and also uses the burrows of other animals. It is primarily nocturnal and crepuscular, skilled at climbing but mainly active on the ground. Its diet consists of rodents, birds, lizards, and other small animals. It is a wild animal of important ecological, scientific, and social value*.

Haloxylon Persicum

Haloxylon persicum is a small tree belonging to the genus *Haloxylon* in the family *Chenopodiaceae*, ranging from 1 to 7 m in height. It thrives in semi-mobile or semi-fixed sand dunes and is a typical constructive species in the desert and under Level-II state priority conservation. It is highly resilient, and tolerant of extreme cold, high temperatures, and drought. Its ecological functions include wind prevention and sand fixation, and its hard wood makes it an excellent source of firewood.

* Included in China's List of Terrestrial Wild Animals of Important Ecological, Scientific, and Social Value.

Chapter 46
Qira Comprehensive Ecological Quality Monitoring Station, Xinjiang (Desert)

Department of Ecology and Environment Monitoring, Ministry of Ecology and Environment of the People's Republic of China and China National Environmental Monitoring Centre, *The Eye on Ecology: China's Ecological Monitoring Records (Volume 1)*, https://doi.org/10.1007/978-981-95-0047-5_46

Location of the Station

Topaaizhik Village, Qira Township, Qira County, Hotan Prefecture, Xinjiang Uygur Autonomous Region.

Lead Organization

Xinjiang Institute of Ecology and Geography, Chinese Academy of Sciences.

Key Monitoring Area

Other key ecosystem service zone.

Latitude and Longitude

80° 44′ E.

37° 1′ N.

Representativeness and Typicality

- Mountain-oasis-desert mixed ecosystem
- Simple ecosystem structure which is subject to multiple stresses, fragile, and difficult to recover
- Extreme atmospheric drought
- Extremely strong wind and sand activity.

Qira Comprehensive Ecological Quality Monitoring Station ("Qira Station" for short) is situated between the Qinghai-Xizang Plateau, demarcated by the Kunlun Mountains, and the Taklamakan Desert, the world's second-largest shifting desert. This region has an extremely harsh natural environment and is one of the primary sources of sandstorms in central Asia, with an average of 20 sandstorm days per year and nearly 200 days of blowing sand and dust haze.

Located on the wind-sand line in the southern margin of the Tarim Basin in southern Xinjiang, it is characterized by basin and range topography that gives rise to typical landscapes including mountain grassland, oasis, and desert ecosystems. The station conducts monitoring of the atmosphere, ecology, surface, and the multi-factorial and multi-process interactions among the atmosphere, water, soil, and biology. It engages in cutting-edge environmental research across multiple fields and disciplines, including the atmosphere, water resources, soil, and vegetation ecology.

Trail around Qira Station

Biodiversity Information

Alhagi Camelorum

It is a sub-shrub herbaceous plant of the genus *Alhagi* in the family *Fabaceae*, and can grow to a height of 25 to 40 cm. In natural conditions, it primarily reproduces asexually through its root system, with seed reproduction being rare. It is a common species in desert areas and is named for being the only forage that camels can eat in the Gobi Desert. Its well-developed, deep root system enables it to absorb groundwater and adapt to the arid desert climate. This plant is widely distributed in desert, Gobi, riverbanks, and farmland edges in Xinjiang, Gansu, and Qinghai, among other regions.

In addition to being used as forage, it is also a valuable medicinal plant and a good source of nectar for bees.

Part V
City Lights

Chapter 47
Beijing Urban Ecosystem Comprehensive Ecological Quality Monitoring Station (Urban Ecosystem)

Department of Ecology and Environment Monitoring, Ministry of Ecology and Environment of the People's Republic of China and China National Environmental Monitoring Centre, *The Eye on Ecology: China's Ecological Monitoring Records (Volume 1)*, https://doi.org/10.1007/978-981-95-0047-5_47

Location of the Station

Changping District, Beijing

Lead Organization

Research Center for Eco-Environmental Sciences, Chinese Academy of Sciences

Key Monitoring Area

Other key ecosystem service zone

Latitude and Longitude

116° 21′ E

40° 1′ N

Representativeness and Typicality

- Typical area with acute urban-regional problems.
- Ecological space for the Beijing-Tianjin-Hebei coordinated development.
- Ecological security shield and preserved water source for Beijing.
- One of the most biodiverse urban ecosystems in the world.

Located in Changping District, Beijing Urban Ecosystem Comprehensive Ecological Quality Monitoring Station ("Beijing Urban Ecosystem Station" for short) primarily research on the evolution of urban ecosystem patterns, their driving mechanisms and ecological effects, as well as urban ecosystem processes. It features a comprehensive monitoring system, including four integrated monitoring platforms (Research Center for of Eco-Environmental Sciences of the Chinese Academy of Sciences, Beijing Teaching Botanical Garden, Mangshan National Forest Park, and Caiyu in Daxing District), one new energy vehicle for mobile monitoring, five long-term biological and soil monitoring sites, 300 long-term biodiversity monitoring points, two groundwater monitoring wells, 14 storm runoff monitoring points, and 51 long-term river/lake water quality monitoring points across different functional zones of city districts.

Mangshan Flux Tower

The station is equipped with over 200 sets of equipment, including meteorological observation systems, soil temperature, humidity, and salinity monitoring systems, plant phenology observation systems, airborne pollen monitoring, wild mammal and bird observation networks, greenhouse gas observation systems, and leaf area meters using images.

Mangshan Mountain

Ardea cinerea; *Turdus merula*; *Passer montanus*; *Aegithalos caudatus*

Phoenicurus auroreus; *Paradoxornis heudei*; *Sinosuthora webbiana*

Prunus sibirica; *Forsythia suspense*

Biodiversity Information

Capreolus Pygargus

Capreolus pygargus, or the Asian roe deer, is included in China's List of Terrestrial Wild Animals of Important Ecological, Scientific, and Social Value, and under Level-II municipal conservation in Beijing. The roe deer has gained huge popularity online due to its unique behavior: when startled, it flares the white fur on its tail, forming a distinctive heart-shaped white rump.

Prionailurus bengalensis

Prionailurus bengalensis is technically the only extant species within the family *Felidae* found in Beijing. It is a top predator in the local food chain and a wild animal under Level-II state priority conservation. Known as the "money cat" in China due to the coin-like spots on its coat, *Prionailurus bengalensis* primarily preys on rodents, rabbits, small birds, and other small animals. It typically inhabits mountainous forests, rural thickets, and forest-edge villages. A couple of years ago, it was spotted at an urban park in Beijing with high human activity, marking the first of such discovery.

Chapter 48
Guangzhou Urban Ecosystem Comprehensive Ecological Quality Monitoring Station, Guangdong Province (Urban Ecosystem)

Department of Ecology and Environment Monitoring, Ministry of Ecology and Environment of the People's Republic of China and China National Environmental Monitoring Centre, *The Eye on Ecology: China's Ecological Monitoring Records (Volume 1)*, https://doi.org/10.1007/978-981-95-0047-5_48

Location of the Station

Haizhu National Wetland Park, Guangzhou City, Guangdong Province.

Lead Organization

Ecological and Environmental Monitoring Center of Guangdong Province.

Participating Organizations

Ecological and Environmental Monitoring Center of Guangzhou.

Guangzhou Haizhu Wetland Center for Research and Education.

South China Institute of Environmental Sciences, Ministry of Ecology and Environment.

Key Monitoring Area

Other key ecosystem service zone.

Latitude and Longitude

113° 20′ E.

23°03′ N.

Representativeness and Typicality

- Typical subtropical urban wetland ecosystem.
- Mixed ecosystem of rivers, inland lakes, and semi-natural orchards in the urban core area of the Pearl River Delta.
- The largest national wetland park in the center of a megacity, known as the "green heart" of Guangzhou.
- High urban biodiversity.

Guangzhou Urban Ecosystem Comprehensive Ecological Quality Monitoring Station ("Guangzhou Urban Ecosystem Station" for short) is located in Haizhu National Wetland Park, known as the "green heart" of Guangzhou. This region has a subtropical monsoon climate with plentiful water resources. Situated at the confluence of the Xijiang River, the Beijiang River, and the Dongjiang River, it hosts numerous rivers and lakes, which give rise to diverse wetland ecosystems.

The monitoring station conducts extensive monitoring and research on the unique urban wetland ecosystems. It is equipped with advanced monitoring devices, including infrared cameras and drone network systems, and has established ecological survey grids, permanent monitoring sites, and an air-space-land integrated ecological quality monitoring system. It has recorded a total of 835 species of vascular plants, 738 species of insects, eight species of amphibians, 22 species of reptiles, 197 species of birds, eight species of mammals, 64 species of fish, and 67 species of benthic animals.

Landscape around Guangzhou Urban Ecosystem Station

Haizhu wetland habitat

Ardea cinerea; *Nycticorax nycticorax*; *Egretta garzetta*; *Psittacula krameri*

Loxura atymnus; *Litchi chinensis*; *Dimocarpus longan*; *Clausena lansium*

Biodiversity Information

Pterorhinus Chinensis

Pterorhinus chinensis belongs to the family *Leiothrichidae* in the order *Passeriformes* and is a wild animal under Level-II state priority conservation. This species measures between 23 and 29 cm in length. It features a black head and throat, an olive-gray belly and undertail coverts, and a distinctive fluffy black forehead with a white edge above it. The bird is predominantly found in Yunnan, Guangxi, and Guangdong in China, as well as various Southeast Asian countries. It is a resident bird in Guangdong. It primarily inhabits the understory and forest edges of tropical rain forests and monsoon forests, as well as bamboo forests, shrubs, and grasslands near agricultural lands and rural settlements.

Hydrophasianus Chirurgus

Hydrophasianus chirurgus belongs to the family *Jacanidae* of the order *Charadriiformes* and is a wild animal under Level-II state priority conservation. It is a relatively large bird measuring between 310 to 580 mm in length. In its summer plumage, it features a white head, chin, throat, and front neck, with a golden-yellow hind neck. The black nape extends into a black stripe along each side of the neck, connecting to the black chest. This species prefers habitats rich in emergent and floating plants, such as freshwater lakes, ponds, and marshes. It is an omnivore, feeding on both aquatic animals and plants. In China, it is mainly distributed throughout the areas south of the 32nd parallel north, including Taiwan, Hainan Island, and southeastern Xizang.

Discovery of New Species

Sphenoraia Haizhuensis

Sphenoraia haizhuensis belongs to the genus *Sphenoraia* in the subfamily *Galerucinae* in the family *Chrysomelidae*. This species was first collected in October 2021 from the core conservation area of Haizhu Wetland using Malaise traps and pitfall traps. The beetle measures 6.0 to 6.5 mm in length and 4.0 to 4.3 mm in width. Its head is prominently convex with distinct punctures, the pronotum is yellow, and there is a comma-shaped or oval black spot on each side of the disc, with seven black spots on each elytron. Analysis of its distribution range reveals that it is a new species endemic to Guangdong.

Falsonnanocerus Haizhuensis

Falsonnanocerus haizhuensis belongs to the genus *Falsonnanocerus* in the tribe *Cnodalonini*, the subfamily *Stenochiinae,* the family *Tenebrionidae*, and the order *Coleoptera*. This species is entirely black in color, measuring 6.2 mm in length, and features a slender elongated oval shape. It has a moderately convex dorsum and a gentle curvature, covered with short yellow hairs. It is a new species that was collected using an insect flight interception trap at Haizhu Wetland Nature School in February 2022.

Chapter 49
Shenzhen Urban Ecosystem Comprehensive Ecological Quality Monitoring Station, Guangdong Province (Urban Ecosystem)

Department of Ecology and Environment Monitoring, Ministry of Ecology and Environment of the People's Republic of China and China National Environmental Monitoring Centre, *The Eye on Ecology: China's Ecological Monitoring Records (Volume 1)*, https://doi.org/10.1007/978-981-95-0047-5_49

Location of the Station

Shenzhen, Guangdong Province.

Lead Organization

Ecological and Environmental Monitoring Center of Shenzhen.

Key Monitoring Area

Guangdong-Hong Kong-Macao Greater Bay Area Biodiversity Conservation Functional Area.

Latitude and Longitude

114° 35′ E.

22° 32′ N.

Representativeness and Typicality

- Planted forest, subtropical evergreen broad-leaved forest, and wetland ecosystems.
- Regionally representative biodiversity and vegetation ecosystems.
- Typical mangrove ecosystem.

Shenzhen Urban Ecosystem Comprehensive Ecological Quality Monitoring Station ("Shenzhen Urban Ecosystem Station" for short) is located in Longgang District, Shenzhen. It focuses on studying the features, drivers, and feedback mechanism of human-nature coupled ecosystems across multiple scales, and developing theoretical frameworks that explain the intricate relationships within human-nature coupled ecosystems and methods for stabilizing these ecosystems. It explores how urbanization and ecological landscapes interact, and engages in urban ecosystem simulation and application.

This station has four locations and is equipped with 2,848 sets and 160 types of equipment, including UAV hyperspectral sensors, environmental DNA (eDNA) monitoring devices, automated pollen observation systems, laser elemental analyzers, LiDAR, and synthetic aperture radar (SAR). The station includes 108 long-term observation points, 45 plant community sample plots, five carbon–water flux observation towers, and the first 20-hectare urban sample plot. The station also houses a well-equipped zoological and botanical specimen room, preserving 12,000 plant specimens, 4,100 insect specimens, 1,100 soil samples, and 150 aquatic organism specimens. Furthermore, four new insect species have been discovered and five species newly recorded.

An aerial view of Shenzhen Urban Ecosystem Station

A panoramic view of the field station

Aethopyga christinae; *Phoenicurus auroreus*

Himantopus himantopus; *Recurvirostra avosetta*; *Megophrys brachykolos*; *Bungarus fasciatus*; *Paramesotriton hongkongensis*; *Viverricula indica*

Biodiversity Information

Platalea Minor

The black-faced spoonbill (*Platalea minor*), a "famed bird" of Shenzhen Bay, is named for its bill resembling *pipa* (a traditional Chinese musical instrument). It is often seen frolicking, foraging, preening, and dancing in the mangrove wetland.

It is under Level-I state priority conservation according to China's List of Wild Animals Under State Priority Conservation published in 2021.

Discovery of New Species

Linan Qiniangmontis

This species was discovered in Qiniangshan, or Qiniang Mountain, in Shenzhen in March 2019, and was published in the international taxonomy journal *Zookeys* in July and identified as a new species. The scape of its antennae is slightly curved at the anterior lateral angle, the fore wings have short veins that gradually taper towards the tip, and the basal segment of the hind wings has a truncate and curved ventral process. In Shenzhen, it is primarily distributed in dry leaf litter layers of mountains at lower elevations or in symbiosis with ants.

Stenus Wutongshanus

This species was discovered in the leaf litter layers of a mixed forest on Wutongshan, or Wutong Mountain, at an elevation of approximately 900 m, in Shenzhen in March 2019. It was officially recognized as a new species when it was published in the international taxonomy journal *Zootaxa* in June of the same year. This species is short-winged, and its body is dark brown with a faint bronze sheen. Each elytron features an orange spot near the lateral margin. Apart from the scape, the antennae, maxillary palps, and legs are yellowish-brown. It prefers to inhabit the shady and humid leaf litter layers of mountainous forest land at higher elevations, with high requirements for environmental quality.

Part VI
Idyllic Countryside

Chapter 50
Hebi Comprehensive Ecological Quality Monitoring Station, Henan Province (Farmland)

Department of Ecology and Environment Monitoring, Ministry of Ecology and Environment of the People's Republic of China and China National Environmental Monitoring Centre, *The Eye on Ecology: China's Ecological Monitoring Records (Volume 1)*, https://doi.org/10.1007/978-981-95-0047-5_50

Location of the Station

Xincun Village, Pangcun Town, Qibin District, Hebi City, Henan Province

Lead Organization

Henan Normal University

Participating Organizations

Chinese Research Academy of Environmental Sciences

Henan Eco-Environmental Technology Center

Henan Geological and Mining Eco-Environmental Technology Co., Ltd.

Hydrological and Water Resources Survey Bureau of Hebi City

Key Monitoring Area

Taihangshan (or Taihang Mountains) Soil Conservation Functional Area

Latitude and Longitude

114° 14′ E

35° 45′ N

Representativeness and Typicality

- Located at the intersection of the Yanshan-Taihangshan region and the Yellow River Basin, a priority area for ecological monitoring.

Hebi Comprehensive Ecological Quality Monitoring Station ("Hebi Station" for short) is located in Xincun Village, Pangcun Town, Qibin District, Hebi City, Henan Province. It primarily monitors and studies the farmland ecosystem within the Yellow River-Haihe River region, as well as forest, wetland, and urban–rural ecosystems. The station engages in coordinated monitoring and research related to climate change, water and sediment dynamics, and biodiversity conservation. The station is equipped with 121 sets of field monitoring equipment and indoor analytical instruments. It consists of a field monitoring room, a biochemical laboratory, a remote sensing information room, and a quality control room, and conducts routine ecosystem monitoring, and has 32 fixed monitoring plots. Rich in biodiversity, this region hosts plant species under Level-I state priority conservation such as *Metasequoia glyptostroboides* and *Ginkgo biloba* and those under Level-II state priority conservation like *Glycine soja.* It is also home to animal species under Level-I state priority conservation including *Ciconia nigra* and *Aythya baeri*, and those under Level-II state priority conservation such as *Prionailurus bengalensis*, *Luscinia calliope*, *Paradoxornis heudei*, and *Falco tinnunculus.*

Yellow River Wetland

Hilly terraces

Metasequoia glyptostroboides; *Ginkgo biloba*; *Glycine soja*; *Ciconia nigra*

Aythya baeri; *Luscinia calliope*; *Falco tinnunculus*

Biodiversity Information

Paradoxornis Heudei

Paradoxornis heudei is a member of the genus *Paradoxornis* in the family *Paradoxornithidae* of the order *Passeriformes*, characterized by its yellowish-brown upper back with black longitudinal stripes. The central tail feathers are sandy brown, while the rest are black with white tips. This bird prefers habitats such as reed thickets along riverbanks and lakeshores, and reed beds, sedge patches, and cattail stands found in estuaries, sandbars, and coastal mudflats. Its diet consists mainly of insects and larvae, including dragonflies, aphids, flies, and scale insects. The breeding season takes place from May to August. It is a rare species endemic to China, known as the "giant panda among birds." The species has persisted for a hundred million years on the planet and can be said a "living fossil" among birds. It is highly dependent on the reed habitat and only survives in suitable habitat conditions. This species is included in China's List of Terrestrial Wild Animals of Important Ecological, Scientific, and Social Value and List of Wild Animals Under State Priority Conservation, and classified as a Near Threatened (NT) species on the IUCN Red List of Threatened Species.

Part VII
Magical Marine Life

Chapter 51
Hangzhou Bay Hekou Comprehensive Ecological Quality Monitoring Station, Zhejiang Province (Marine Ecosystem)

Department of Ecology and Environment Monitoring, Ministry of Ecology and Environment of the People's Republic of China and China National Environmental Monitoring Centre, *The Eye on Ecology: China's Ecological Monitoring Records (Volume 1)*, https://doi.org/10.1007/978-981-95-0047-5_51

Location of the Station

Zhoushan City, Zhejiang Province

Lead Organization

Ecological and Environmental Monitoring Center of Zhejiang Province

Key Monitoring Area

Key estuary ecosystem conservation and restoration area within the Yangtze River Delta

Latitude and Longitude

122° 12′ E

29° 59′ N

Representativeness and Typicality

- A trumpet-shaped tidal bay as the Qiantang River meets the East China Sea
- Unique freshwater estuary-marine ecosystem rich in rare and endangered species
- Area of marine ecological importance
- The Zhoushan fishing area, a key location for fisheries (feeding ground, spawning ground, wintering ground, and migration route).

Hangzhou Bay Hekou Comprehensive Ecological Quality Monitoring Station ("Hangzhou Bay Hekou Station" for short) is located in Zhoushan City, Zhejiang Province. The Hangzhou Bay is distinguished by its unique freshwater estuary-marine ecosystem and diverse types of ecosystems. The primary goal of this station is to promptly detect ecological disturbances in the Hangzhou Bay estuary, collect monitoring data on estuary ecosystems, and support the conservation and restoration of significant estuary ecosystems in the Yangtze River Delta. The station is equipped with 481 sets of analytical equipment, capable of monitoring 533 parameters in 14 categories across both land and sea environments. It possesses a 1,000-ton-class vessel for eco-environmental monitoring (China Huan Jian Zhe 001), as well as drones and automated buoy stations. It has 43 monitoring sites and conducts annual multi-element monitoring over approximately 29,400 square kilometers of sea.

Eco-environmental monitoring vessel (China Huan Jian Zhe 001)

Wuzhishan Bird Island

Thalasseus bernsteini; *Rapana venosa*; *Philyra carinata*

Thalasseus bergii

Biodiversity Information

Solea Ovata

Solea ovata has an oval-shaped body and primarily inhabits tropical marine environments, often found in estuaries as well. This species typically rests on the seafloor and buries most of its body in sand or mud, leaving only its eyes exposed to observe the surroundings. This species can subtly change its body color to blend with the environment, using an ambush strategy to capture small benthic invertebrates that come close. As a carnivorous species, it is not good at swimming and usually moves by undulating its body in a wavelike motion.

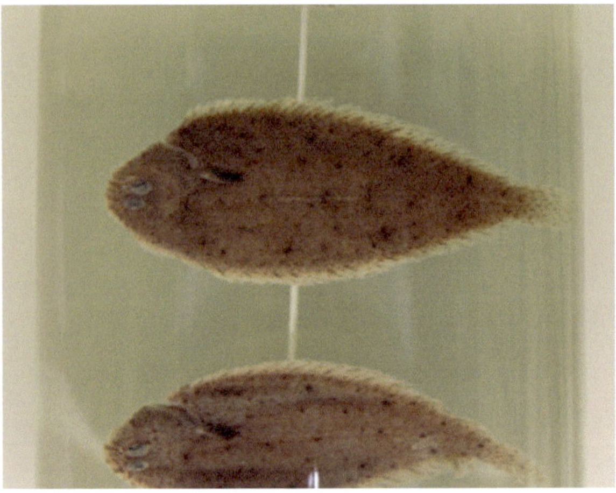

Sepiella Maindroni

Sepiella maindroni is a cuttlefish belonging to the family *Sepiidae* of the order *Cephalopoda*. It has a shield-shaped trunk, with a body length that is twice its body width. The back of the trunk is adorned with numerous nearly elliptical white spots. It is distributed in the coastal waters of Zhejiang and Fujian. The cuttlefish exhibits positive phototaxis, preferring dim light and orienting away from bright light. It is highly valued for its culinary and medicinal properties. It was one of the top four marine aquaculture produces in China during the 1970s and 1980s. The Zhoushan sea area is one of the primary cuttlefish regions, but overfishing has led to a dramatic decline in its population, making it rare in these waters. Artificial breeding and hatchery methods have been adopted in recent years to help recover its populations.

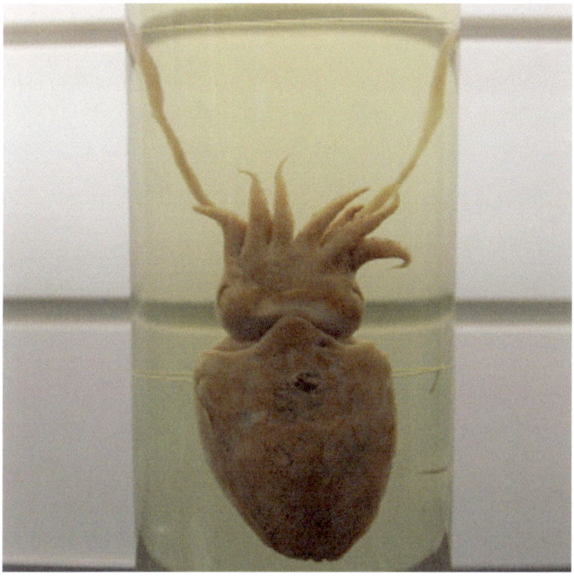

Chapter 52
Jiaozhou Bay Comprehensive Ecological Quality Monitoring Station, Shandong Province (Marine Ecosystem)

Department of Ecology and Environment Monitoring, Ministry of Ecology and Environment of the People's Republic of China and China National Environmental Monitoring Centre, *The Eye on Ecology: China's Ecological Monitoring Records (Volume 1)*, https://doi.org/10.1007/978-981-95-0047-5_52

Location of the Station

Qingdao City, Shandong Province.

Lead Organization

Institute of Oceanology, Chinese Academy of Sciences.

Latitude and Longitude

120° 12′ E.

36° 3′ N.

Representativeness and Typicality

- Temperate semi-enclosed bay subject to the impact of climate change and human activity.
- A witness to the development of Qingdao, a port city known for its beautiful beaches.
- A healthy, sustainable bay amid the rapid development of coastal regions.

Jiaozhou Bay Comprehensive Ecological Quality Monitoring Station ("Jiaozhou Bay Station") is located in Qingdao, Shandong Province. Jiaozhou Bay covers an area of 374.4 square kilometers, stretching 28 km from east to west and 32 km from north to south. It has a coastline of over 30 km, with tidal flats covering 56,000 *mu* (3,733.33 hectares) and shallow sea areas spanning nearly 870,000 *mu* (5,800 hectares). Jiaozhou Bay is a typical bay in China that combines the functions of ports, tourism, and aquaculture. The station focuses on monitoring ecosystems in the Jiaozhou Bay and its adjacent waters through broad-scale monitoring, targeted regional monitoring, and coordinated monitoring of river-estuary-bay ecosystems. Using a combination of traditional monitoring, intensive monitoring, and automated monitoring techniques, the station conducts systematic research on long-term ecosystem changes, evolution, drive mechanisms, trends, and mitigation strategies, with the aim to identify the key processes for maintaining the health and sustainability of bay ecosystems.

Jiaozhou Bay research ship named "Chuangxin" (meaning "innovation")

An aerial view of Jiaozhou Bay Station and surrounding areas

Aquaculture farming in the estuary

Aquaculture farming in the observation area; Floating ice in winter

Estuarine wetland

Biodiversity Information

Calanus Sinicus

Calanus sinicus is a dominant zooplankton species in Jiaozhou Bay, primarily distributed in China's coastal waters and along the Japanese coast. It has a broad distribution and large population, and plays a vital role in the marine food web. It is an important food source for many economically important fish species, and its population dynamics directly influence the fishery resources.

Chapter 53
Beibu Bay Comprehensive Ecological Quality Monitoring Station, Guangxi (Marine Ecosystem)

Department of Ecology and Environment Monitoring, Ministry of Ecology and Environment of the People's Republic of China and China National Environmental Monitoring Centre, *The Eye on Ecology: China's Ecological Monitoring Records (Volume 1)*, https://doi.org/10.1007/978-981-95-0047-5_53

Location of the Station

Shatian Town, Hepu County, Beihai City, Guangxi Zhuang Autonomous Region.

Lead Organization

Marine Environment Monitoring Center of Guangxi.

Participating Organizations

National Marine Environmental Monitoring Center.

Guangxi Hepu Rugen National Nature Reserve Administration.

Key Monitoring Area

Beibu Bay Coastal Wetland Ecosystem Conservation and Restoration Area.

Latitude and Longitude

109° 39′ E.

21° 31′ N.

Representativeness and Typicality

- Marine ecosystem with subtropical bay and tropical characteristics.
- Key area along the New International Land-Sea Trade Corridor.
- Typical later comer in bay economy and maritime economy.
- Representative of eco-environmental protection initiatives across provinces and countries and in cooperation with ASEAN.

Beibu Bay Comprehensive Ecological Quality Monitoring Station ("Beibu Bay Station" for short) is located in Hepu County, Beihai City, and houses a variety of marine ecosystems, including mangroves, seagrass beds, coral reefs, seaweed fields, and estuarine bays. Notably, the diversity and preservation of mangrove species are among the best in the country, and the seagrass beds in Guangxi constitute the primary region for the distribution of subtropical seagrass species in China. The region is rich in marine biodiversity, with about 1,587 species of marine organisms having been recorded. A total of 138 species are under state priority conservation, including *Balaenoptera edeni*, *Sousa chinensis*, *Dugong dugon*, and *Tachypleus tridentatus*. The Beibu Bay is home to China's third-largest known *Sousa chinensis* community, and is also the only coastal region in China where large baleen whales have been frequently observed throughout the year.

The station focuses on monitoring typical marine ecosystems, conducting trend observations of coastal ecology, and providing early warnings for marine ecological disasters. It is equipped with over 600 devices including X/L-band satellite receiving systems, inductively coupled plasma mass spectrometry (ICP-MS), vehicle-mounted gas chromatographs, and portable emergency monitoring devices. It conducts air-space-land-sea integrated ecological quality monitoring, including satellite, drone, and ground-based monitoring.

Patrol vessel for ecological monitoring

Coral reef ecosystem

Neophocaena phocaenoides; *Sousa chinensis*

Tachypleus tridentatus; *Hydrophasianus chirurgus*; *Tringa guttifer*; *Platalea minor*; *Merops viridis*

Biodiversity Information

Balaenoptera Edeni

Balaenoptera edeni belongs to the genus *Balaenoptera* in the family *Balaenopteridae* of the suborder *Mysticeti* and the order *Cetacea.* It is a wild animal under Level-I state priority conservation in China. With an extremely limited population, it is considered one of the most mysterious and precious members in the whale family, and is included in the IUCN Red List of Threatened Species. From around the Mid-Autumn Festival to the end of April the following year, *Balaenoptera edeni* seasonally appears near Weizhou Island. They are the first large whale population to be frequently observed in China's coastal waters since the 1980s. *Balaenoptera edeni* has a dark gray color and a yellowish-white segmental venter, measures about 12 m long on average, and features stomata on the top of the head. It is a rare and valuable mammal in marine ecosystems and serves as an important indicator species. As of 2023, a total of 60 individuals had been identified in the waters around Weizhou Island.

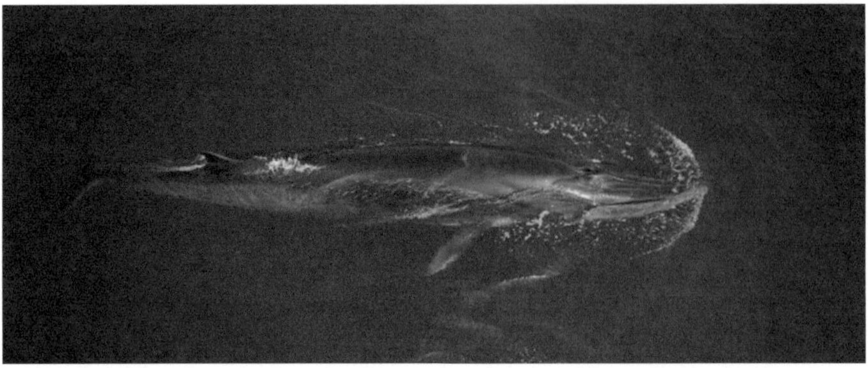

Mangrove

Mangroves are a unique wetland woody plant community found in tropical and subtropical coastal and estuarine intertidal zones. Known as the "guardians of coastlines.," they play important roles in wind and wave mitigation, coastal protection, farmland preservation, and reduction of saltwater intrusion. Mangroves in Guangxi are primarily distributed in Tieshan Port and Lianzhou Bay in Beihai City, Maowei Sea in Qinzhou City, Fangchenggang Bay and Zhenzhu Bay in Fangchenggang City, and other estuary bays. Guangxi is home to diverse mangrove species, hosting 12 true mangrove species (including two exotic species), which account for 44 percent of the country's total. Additionally, there are eight semi-mangrove species. The mangrove ecosystems in the region are varied, including the largest urban mangrove and sand mangrove forests in Jinhai Bay of Beihai City, the largest contiguous natural *Rhizophora stylosa* forest in Yingluo Port in Shankou Town of Hepu County, the unique island-cluster mangrove forest in Qinzhou Port Qishierjing scenic area, and the typical estuarine mangrove forest in Zhenzhu Bay, Fangchenggang.

1. *Balaenoptera edeni* monitoring and survey
2. *Balaenoptera edeni* feeds near Weizhou Island
3. *Balaenoptera edeni* swims near Weizhou Isand
4. Mangroves

Seagrass

Seagrass is the only higher angiosperm that thrives in seawater environments. It is widely distributed in coastal waters of tropical and temperate regions around the world. Seagrasses typically grow in shallow waters and estuary areas in intertidal zones and shallow subtidal zones. China currently has 22 species of seagrass, belonging to 10 genera in four families. Seagrasses in Guangxi are mainly distributed in Hepu County, Beihai City, and Zhenzhu Bay, Fangchenggang City. The seagrass beds in Hepu are located in the Beihai Hepu Rugen National Nature Reserve and its surrounding waters, which host such species as *Halophila ovalis* and *Halophila beccarii.* The seagrass beds in Zhenzhu Bay are located in the northern part of Jiaodong (Zhenzhu Bay), Fangchenggang City, and the key species include *Halophila beccarii* and *Zostera japonica.*

Halophila ovalis

Seagrass bed ecosystem

Discovery of New Species

Elysia Leucolegnote

Elysia leucolegnote, a newly recorded species found in Beihai Wetland Park, belongs to the order *Sacoglossa* in the class *Gastropoda* of the phylum *Mollusca*. It is a small benthic animal that resembles "walking green leaves" and commonly inhabits the intertidal zones of mangrove forests. This species consumes algae and integrates the chloroplasts from the algae into its own cells to perform photosynthesis. It is currently the only known model organism among complex multicellular organisms that can maintain active chloroplasts within their cells. The discovery of this species has updated the biodiversity records of mangrove ecosystems.

Phoca Largha

Phoca largha is a newly recorded marine mammal discovered in the waters around Weizhou Island and the northern coast of Beihai City.

It belongs to the genus *Phoca* in the family *Phocidae* of the order *Pinnipedia.* Adults are typically grayish-yellow or dark gray, with a light yellow belly and numerous dark spots on their back. This species is primarily distributed in the higher-latitude cold regions of the northwestern Pacific. It's uncommon for the species to be found in the tropical-subtropical waters of Beibu Bay.

Chapter 54
Hainan Coral Reef Comprehensive Ecological Quality Monitoring Station (Marine Ecosystem)

Department of Ecology and Environment Monitoring, Ministry of Ecology and Environment of the People's Republic of China and China National Environmental Monitoring Centre, *The Eye on Ecology: China's Ecological Monitoring Records (Volume 1)*, https://doi.org/10.1007/978-981-95-0047-5_54

Location of the Station

Haikou City, Hainan Province; Luhuitou Peninsula, Sanya City, Hainan Province.

Lead Organization

Ecological and Environmental Monitoring Center of Hainan Province.

Participating Organization

South China Sea Institute of Oceanology, Chinese Academy of Sciences.

Key Monitoring Area

Hainan Island coastal zone.

Latitude and Longitude

109° 28′ E.

18° 12′ N.

Representativeness and Typicality

- Typical marine ecosystems such as coral reefs, mangroves, and seagrass beds.
- Coral reefs in this region cover about 95 percent of the country's total.
- Covering Sanya Coral Reef National Nature Reserve and Tongguling National Nature Reserve.

Hainan Coral Reef Comprehensive Ecological Quality Monitoring Station ("Hainan Coral Reef Station" for short) is located on Luhuitou Peninsula in Sanya City. This area has a tropical monsoon climate and is known as a "natural greenhouse."

This station is the only monitoring station in China that features coral reef monitoring, a main part of its comprehensive marine monitoring activities. It monitors various elements such as seawater quality, marine organisms, and marine sediments. The station has well-equipped infrastructure and high-resolution, high-sensitivity coral reef monitoring devices, including conductivity-temperature-depth (CTD) profilers, real-time underwater observation systems, fully automated water quality analyzers, and total dissolved inorganic carbon analyzers. It also has a 625-square-meter observation field and a 3.308-hectare coral reef experimental plot. Surveys find that the waters off the east coast of Hainan Island are home to 136 species of reef-building hard corals across 33 genera of 14 families, and the common species include *Porites lutea*, *Galaxea fascicularis*, and *Acropora hyacinthus*. The waters off the west coast of Hainan Island host 58 species of reef-building hard corals across 19 genera of nine families, and the common species include *Porites lobata*, *Porites lutea*, and *Favites complanata*.

Coral nursery frames; Coral nursery trees

Biodiversity Information

Galaxea Fascicularis

Galaxea fascicularis is a species of colonial stony coral in the genus *Galaxea*, family *Euphylliidae*, order *Scleractinia*, class *Hexacorallia*, subphylum *Anthozoa*, phylum *Cnidaria*, and kingdom *Animalia*. Due to the pressure from crowded tentacles, the corallites, or skeletal cups, can have several different shapes, such as circular, oval, rectangular, or even irregular. A corallite measures 6 to 10 mm in diameter. There are numerous, densely packed corallites, contributing to the formation of a mound-shaped skeletal structure. In colonial life, the corals typically form flat, plate-like, or branching structures.

Acropora Valida

The coral forms dense corymbose colonies, varying from bushy to table-shaped. The branches have a uniform thickness and are robust. The tips of the living coral are light yellow, while the middle and lower parts are brown or violet.

Chapter 55
Xisha Tropical Island Comprehensive Ecological Quality Monitoring Station, Hainan Province (Marine Ecosystem)

Department of Ecology and Environment Monitoring, Ministry of Ecology and Environment of the People's Republic of China and China National Environmental Monitoring Centre, *The Eye on Ecology: China's Ecological Monitoring Records (Volume 1)*, https://doi.org/10.1007/978-981-95-0047-5_55

Location of the Station

Yongxing Island, Sansha City, Hainan Province.

Lead Organization

Ecological and Environmental Monitoring Center of Hainan Province.

Key Monitoring Area

Hainan Island coastal zone.

Latitude and Longitude

112° 16′ E.

16° 58′ N.

Representativeness and Typicality

- The Xisha Islands are characterized by sea-and-land duality, unique resources, and ecological fragility.
- Primary area of tropical islands in China.
- Important area of typical ecosystems such as coral reefs.
- Important feeding ground, spawning ground, wintering ground, and migration route for commercially important fish in the South China Sea.

Xisha Tropical Island Comprehensive Ecological Quality Monitoring Station ("Xisha Station" for short) is located in Sansha City, Hainan Province, characterized by a tropical marine monsoon climate. The landforms of Xisha Islands are all represented by coral reefs, which are classified into islands, cays, reefs, submerged banks, and submerged shoals based on their elevation relative to sea level.

The station is well-equipped with instruments such as UV spectrophotometers, fluorescence microscopes, stereoscopes, drones, and in-situ online video monitoring systems for coral reefs. The station is capable of monitoring various elements such as coral reefs, seagrass beds, seawater quality, marine organisms, marine sediments, air quality, as well as soils and vegetation on islands and reefs. Currently, 147 species of reef-building hard corals from 30 genera of 10 families have been identified in this area. The common species include *Acropora pulchra*, *Porites lutea*, *Plesiastrea versipora*, and *Platygyra carnosus*. Additionally, continuous monitoring is conducted on the ecological environment of green turtle nesting grounds and the dynamics of their breeding populations.

Yagong Island

Coral reef monitoring in Xisha Station; Jinqing Island; Nanshazhou

Porites lutea; *Pocillopora meandrina*

Close-up of coral reef; *Montipora samarensis*

Biodiversity Information

Cephalopholis Argus

The body of *Cephalopholis argus* is typically dark brown or purplish-black, and its sides and fins are marked with gray spots encircled by black rings. This species can change its coloring, and its diet consists of small fish and invertebrates. It is found in the Indo-Pacific region.

Acropora Hyacinthus

Acropora hyacinthus builds a disc-shaped skeleton, forming corymbose colonies. It encrusts on live or dead coral substrates, featuring interlocking, short, and robust branches. Living corals are either monochromatic (brownish-yellow, brownish, coffee-brown) or multicolored (deep coffee-brown with green).

MIX
Papier aus verantwortungsvollen Quellen
Paper from responsible sources
FSC® C105338

If you have any concerns about our products,
you can contact us on
ProductSafety@springernature.com

In case Publisher is established outside the EU,
the EU authorized representative is:
Springer Nature Customer Service Center GmbH
Europaplatz 3, 69115 Heidelberg, Germany

Printed by Libri Plureos GmbH
in Hamburg, Germany